Lecture Notes in Mathematics   2093

Raphael Kruse

Strong and Weak Approximation of Semilinear Stochastic Evolution Equations

 Springer

Raphael Kruse
Department of Mathematics
Bielefeld University
Bielefeld, Germany

ISBN 978-3-319-02230-7   ISBN 978-3-319-02231-4 (eBook)
DOI 10.1007/978-3-319-02231-4
Springer Cham Heidelberg New York Dordrecht London

Lecture Notes in Mathematics ISSN print edition: 0075-8434
ISSN electronic edition: 1617-9692

Library of Congress Control Number: 2013952955

Mathematics Subject Classification (2010): 65C30, 60H15, 65M60, 60H07, 35B65

Springer is part of Springer Science+Business Media (www.springer.com)

To my love Christine

Preface

This monograph grew out of my Ph.D. thesis, which I wrote at Bielefeld University between 2008 and 2012. Its main objective is the analysis of numerical methods, which approximate solutions to stochastic evolution equations (SEEq) on infinite dimensional state spaces. I therefore invite the reader to join me at the meeting point of several fascinating mathematical fields: numerical analysis, probability theory and stochastic analysis, PDE theory, and functional analysis.

For me the study of numerical methods always has the convenient advantage that I can investigate my object of interest not only theoretically but also during computer experiments. In moments of doubts, a numerical simulation may motivate me to look harder for a theoretical proof or it helps avoid wasting time on a dead end if I am on the wrong track. Computer experiments also serve as a constant well of inspiration by making me stumbling over unexpected phenomena, whose understanding often raises interesting theoretical questions.

To this respect numerical methods for SEEqs are a very grateful study object. Even without changing the initial parameters one may be able to see many different kinds behaviors just by the presence of the random forcing terms. A very interesting task is then to puzzle out which phenomena are due to the underlying stochastic evolution equation and which stems from the application of a numerical method and their inevitable errors.

For this it is of course necessary to first have a closer look at stochastic evolution equations and their properties. As this book is mainly written from the viewpoint of a numerical analyst, we concentrate on basic questions which are important for the design and the behavior of numerical schemes. In particular, we study existence, uniqueness, and spatio-temporal regularity of solutions to SEEqs.

After having made ourselves familiar with the object which we want to approximate, we next have to think about the aim of the approximation. Shall our numerical scheme give a good pathwise approximation of the solution or shall it just closely reproduce certain statistical properties? The two possibilities are related to the two different kinds of errors which we study in this monograph.

On the one hand, there is the so-called *strong error of convergence*, which is measured in terms of the L^p-norm for random variables. This error is small if for a given initial state and a given sample path of the forcing random noise the corresponding trajectories of the exact solution and the numerical scheme are close to each other.

On the other hand, the *weak error of convergence* is already small if the laws of the exact solution and the numerical scheme almost coincide. Usually, a strongly convergent scheme is also weakly convergent and the strong order of convergence is a lower bound for the weak order of convergence. However, a well-known rule of thumb states that for many numerical schemes the weak order is actually twice the strong order of convergence.

While the analysis of the strong error often relies on similar techniques as for the deterministic problem, the weak error analysis is often much more demanding and usually builds upon the solution to Kolmogorov's equation associated with the SEEq. Consequently, the theory concerning the weak error analysis is by far not so mature and much more restrictive as for the strong error.

To some degree the same holds true for this monograph: While the strong error analysis is carried out for more general SEEqs with multiplicative noise, we only consider linear SEEqs with additive noise in the discussion of the weak error. But for the latter, my main intention is to introduce the reader to the beautiful theory of the Malliavin Calculus. I hope that the introduction to this topic is particularly accessible to readers who have already gained some familiarity with Wiener processes and appreciate a more functional analytical approach as in the standard literature.

The Malliavin Calculus is then used to open a new path to the weak error analysis, which completely bypasses Kolmogorov's equation. This new approach enjoys several advantages as, for example, the possibility to consider the weak error for SEEqs with non-autonomous stochastic coefficients. But concerning the treatment of equations with multiplicative noise much more research needs to be done.

I tried to do my best in order to keep this monograph mostly self-contained. However, I did not succeed everywhere: At the very least the reader should have a basic knowledge of standard techniques in functional analysis. It is also helpful if the reader is familiar with numerical methods for finite dimensional stochastic differential equations. More advanced topics such as Wiener processes on Hilbert spaces, strongly continuous semigroups, and Galerkin finite elements are provided without proof.

Notation

In order not to hide easily understandable ideas behind a complicated notation, I tried to keep everything as standard and simple as possible. However, since these notes are written in the intersection of several different mathematical fields, some collisions were inevitable.

For example, in stochastic analysis it is very common to denote the time evaluation of a stochastic process by an index, that is $X(t) = X_t$. However, this habit collides with the notation in numerical analysis, where the spatial discretization of a function u is usually denoted by u_h. From my point of view, it generates much less confusion if I stick to the classical notation in analysis to write $X(t)$ for the time evaluation of a stochastic process and I do not change the notation for the numerical methods to something nonstandard. Please accept my apologies in advance if this is the cause of new confusion which I was not able to foresee.

Further, throughout this book I apply the concept of *generic constants*. Usually denoted by the letter C, I understand by a generic constant a positive and finite entity, which may vary from occurrence to occurrence and depends on exogenous parameters such as the final time T, the Lipschitz constants of the nonlinearities f and g, or the norm of the initial value X_0. However, C is always chosen in a way that it is independent of endogenous parameters of the numerical method such as the spatial and temporal step sizes h and k.

By this means, perhaps interesting relationships between the exogenous parameters may remain hidden and it is unclear, in which way the constant C behaves, if exogenous parameters are varied. On the other hand, the derivation of explicit representations of all constants would have complicated many proofs and possibly distracted the reader from the underlying ideas and techniques.

Otherwise, new notation is usually explained at first appearance or found in the symbol list.

Acknowledgments

This text would not have been possible without the invaluable support and guidance of so many people.

First of all, I would like to mention my Ph.D. supervisor Prof. W.-J. Beyn (Bielefeld University). I profited very much from his experience and his ability to ask the right questions. He always gave me the freedom to follow my own ideas, but he also did a very good job in preventing me from getting lost in the process.

I also would like to express my sincere gratitude to Prof. S. Larsson (Chalmers University of Technology), who was my patient guide into the finite element world. But perhaps even more importantly, he broadened my horizons in many aspects of life, not just mathematics.

Another constant source of inspiration was Prof. M. Röckner (Bielefeld University), who taught me the fundamentals of stochastic analysis and always knew a precise reference to someone who already solved the problems I just stumbled over.

There are many more colleagues whom I would like to thank for their helpful and interesting discussions, advice, and support. In particular, I mention Prof. A. Jentzen (ETH Zürich), who gave many helpful comments to several aspects of these notes. Many of my results proceed on the basis of his work. Further, I would like to mention A. Andersson and F. Lindgren (both Chalmers University of Technology) who inspired my interest in Malliavin Calculus and weak convergence for SPDEs.

I also say thank you to Dr. S. Cox, D. Holtz and Dr. T. Hüls for helping me to track down an uncountable number of typos. Without their help the manuscript would be much worse as it is now. The same is true for D. Otten who supported me in my teaching duties, especially during the last weeks of writing.

Most importantly, I would like to thank my girlfriend Christine, my family, and all my friends and colleagues for their encouragement, backing, and understanding, in particular during the time when I was too consumed by my own work. I hope to get the chance to return the favor.

Last but not least, I would like to thank an anonymous referee, whose comments and insights were very helpful in improving [50], and consequently, also the results in Sects. 2.5 and 2.6. His comments also strongly helped to improve the classification of these results into the existing literature in Sect. 1.1.

Most of the research in Sects. 2.5 and 2.6 and Chap. 3 has been carried out during an extended stay at Chalmers University of Technology, Gothenburg, Sweden. I would like to thank my host Prof. S. Larsson and his working group for the generous hospitality and the inspiring environment they created for me. In this matter, my thanks also go to Prof. W.-J. Beyn and the German Academic Research Service (DAAD) for making this wonderful stay possible.

Since my Ph.D. thesis was financially supported by the CRC 701 "Spectral Structures and Topological Methods in Mathematics" and the DFG-IGK 1132 "Stochastics and Real World Models," the same holds true for this monograph.

Bielefeld, Germany and Zürich, Switzerland   Raphael Kruse
August 2013

Contents

Symbols

a.s.	Almost surely
EOC	Experimental order of convergence, p. 137
SEEq	Stochastic evolution equation, p. 20
SPDE	Stochastic partial differential equation, p. 22
SODE	Stochastic ordinary differential equation, p. 25
w.r.t.	With respect to
$\mathbb{N}$	Set of all positive integers $1, 2, 3, \ldots$
$\mathbb{R}$	Set of all real numbers
T	A positive real number, usually denoting the upper bound of an interval $[0, T]$
$s \wedge t$	Minimum of $s, t \in \mathbb{R}$
$s \vee t$	Maximum of $s, t \in \mathbb{R}$
$\delta_{i,j}$	Kronecker-δ: that is $\delta_{i,j} = 1$ if $i = j$, else $\delta_{i,j} = 0$
U, H	Separable Hilbert spaces
H'	Dual space of H
$(\cdot, \cdot)_U$	Inner product of a given Hilbert space U
$(\cdot, \cdot)$	Inner product of the Hilbert space H
U_0	Hilbert space $Q^{\frac{1}{2}}(U)$, p. 14
B	A Banach space
$\mathrm{dom}(A)$	Domain of an operator A
$C([0, T]; B)$	Space of all continuous mappings $f : [0, T] \to B$
$C^k(H_1; H_2)$	Space of all continuous mappings $f : H_1 \to H_2$, which are k-times continuously Fréchet-differentiable
$C_p^k(H_1; H_2)$	Space of all $f \in C^k(H_1; H_2)$, such that f and all its derivatives are at most polynomially growing
$C_b^k(H_1; H_2)$	Space of all $f \in C^k(H_1; H_2)$, such that f and all its derivatives are bounded
$\mathbb{D}^{1,p}(H)$	Space of all p-fold integrable H-valued random variables which are differentiable w.r.t. Malliavin derivative, p. 92
$(S, \mathscr{G}, \mu)$	A measure space on a set S with σ-field $\mathscr{G}$ and measure μ

$(\Omega, \mathscr{F}, \mathbf{P})$	Probability space
$\mathbf{E}[Z]$	$\mathbf{E}[Z] = \int_\Omega Z(\omega)\, d P(\omega)$, expectation of Z
$\mathscr{B}(S)$	Borel-σ-field of S
$\mathscr{P}_T$	σ-field of predictable stochastic processes, p. 17
$dt, dx, d\sigma$	Lebesgue measure with respect to integration variables t, x, σ
$H^k(0, 1)$	Sobolev space of square-integrable functions on $[0, 1]$ with existing weak derivatives up to order $k \in \mathbb{N}$
$H_0^1(0, 1)$	Sobolev space of square-integrable functions on $[0, 1]$ with existing first order weak derivative and zero boundary conditions
$\dot{H}^s$	$\dot{H}^s = \mathrm{dom}(A^{\frac{s}{2}})$, domain of $A^{\frac{s}{2}}$, p. 162
$\mathscr{H}_p$	Space of predictable stochastic processes, p-fold integrable, p. 27
$L^p(S, \mathscr{G}, \mu; B)$	Space of all p-fold integrable mappings $f : S \to B$, $p \geq 1$
$L^p(S; B)$	$L^p(S; B) = L^p(S, \mathscr{G}, \mu; B)$, if $\mathscr{G}$ and μ are known
$L(U_1, U_2)$	Space of bounded linear operators from U_1 to U_2
$L(U)$	$L(U) = L(U, U)$
$L_2(U_1, U_2)$	Hilbert space of all Hilbert–Schmidt operators from U_1 to U_2, p. 16
$L_2(U)$	$L_2(U) = L_2(U, U)$
L_2^0	$L_2^0 = L_2(U_0, H)$, p. 16
$L_{2,r}^0$	$L_{2,r}^0 = L_2(U_0, \dot{H}^r)$, p. 22
$\mathscr{N}_W^2(0, T; H)$	Set of all stochastic processes, Itô-integrable w.r.t. W, p. 17
$\mathscr{S}(H)$	Set of smooth and cylindrical random variables with values in H, p. 86
$\|\cdot\|_B$	Norm of a given Banach space B
$\|\cdot\|$	Norm of the Hilbert space H or its induced operator norm in $L(H)$
$\|\cdot\|_s$	Norm of the Hilbert space $\dot{H}^s$, pp. 20, 162
$\|\cdot\|_T$	Norm on $\mathscr{N}_W^2(0, T; H)$, p. 17
$\langle \cdot, \cdot \rangle$	Duality pairing between $\dot{H}^{-1}$ and $\dot{H}^1$, p. 55
A^*	Adjoint of an operator A
Id_U	Identity operator $\mathrm{Id}_U : U \to U$ in $L(U)$
1_A	Indicator function for the measurable set A
I_h	Interpolation operator, p. 130
J_h	Finitely based interpolation operator, p. 135
R_h	Ritz projector, p. 54
$E(t)$	Semigroup generated by $-A$, pp. 20, 159
$h \otimes u$	Tensor product of $h \in H$ and $u \in U$, p. 86
$\mathrm{Tr}(Q)$	Trace of $Q \in L(U)$, p. 11
$(\beta_j)_{j \in \mathbb{N}}$	A family of independent, real-valued Brownian motions,
W	Wiener process, p. 12

Chapter 1
Introduction

In this chapter we familiarize ourselves with semilinear stochastic evolution equations, which are the starting point of this monograph. Then we present our main results and give a glimpse into the most important methods and techniques which are used in this text.

Let H be a separable Hilbert space and let the linear operator $A: \mathrm{dom}(A) \subset H \to H$ be positive definite, self-adjoint with compact inverse. For simplicity, we assume throughout this chapter that H consists of all square-integrable real-valued mappings on the unit interval, that is $H = L^2([0,1], \mathscr{B}([0,1]), \mathrm{d}x; \mathbb{R})$, and that $-A$ denotes the Laplace operator on $[0,1]$ with Dirichlet boundary conditions. Consequently, $-A$ is the generator of an analytic semigroup of contractions $(E(t))_{t \in [0,T]}$.

By $(\Omega, \mathscr{F}, \mathbf{P})$ we denote a complete probability space with filtration $(\mathscr{F}_t)_{t \in [0,T]}$. Then, we consider a semilinear stochastic evolution equation of the form

$$\mathrm{d}X(t) + \big[AX(t) + f(t, X(t))\big] \mathrm{d}t = g(t, X(t)) \, \mathrm{d}W(t), \quad \text{for } 0 \le t \le T,$$
$$X(0) = X_0. \tag{1.1}$$

Here, the solution $X: [0,T] \times \Omega \to H$ is a Hilbert space valued stochastic process, W denotes an H-valued Wiener process and the nonlinearities $f: [0,T] \times \Omega \times H \to H$ and $g: [0,T] \times \Omega \times H \to L(H)$ satisfy certain Lipschitz conditions. In order to keep this chapter as focused on our main results as possible, all details concerning the assumptions on f, g, W and X_0 are deferred to Sect. 2.3.

We treat (1.1) in terms of the semigroup approach as it is found in the book [18, Chap. 7] by Da Prato and Zabczyk. That is, we understand X as a *mild solution* (see Definition 2.18) and we also apply this methodology to our numerical approximation schemes.

Further, let a nonlinear functional $\varphi: H \to \mathbb{R}$ be given, which we always assume to be as smooth as needed. Our primary task, which we want to accomplish in this book, is the numerical approximation of the real number

$$\mathbf{E}\big[\varphi(X(T))\big]. \tag{1.2}$$

R. Kruse, *Strong and Weak Approximation of Semilinear Stochastic Evolution Equations*, Lecture Notes in Mathematics 2093, DOI 10.1007/978-3-319-02231-4_1, © Springer International Publishing Switzerland 2014

In this monograph we omit all questions concerning the computation of the expectation by Monte Carlo methods and we focus on the discretization of the solution $X(T)$ itself. For a more computational treatment of Monte Carlo algorithm we refer to [29] and references therein.

If one wants to write a computer program which computes an approximation of (1.2), by the discrete and finite nature of the computer's memory, it is inevitable to represent the solution X on finite partitions in space and time. In this monograph we are therefore concerned with discretizations of the time interval $[0, T]$ and the Hilbert space H.

In addition, for the implementation of the Wiener noise W one also needs to consider a discretization of its covariance operator. However, this question is not addressed in this book and we refer to [3, Sect. 4] and the references therein.

As it is common in numerical analysis of deterministic partial differential equations, we first analyze the so called spatially semidiscrete approximation of (1.1), that is, we only discretize with respect to the Hilbert space H. For this we apply the well-established theory of Galerkin finite element methods from [66]. This theory has the advantage that it is suitable to treat different Galerkin methods in an unified setting.

In particular, let us stress that the theory of Galerkin finite element methods does not use any explicit knowledge of the eigenfunctions and eigenvalues of the operator A. This important feature is inherited by our results which are therefore applicable in a variety of situations.

To be more precise, let us denote by $(S_h)_{h\in(0,1]}$ a family of finite dimensional subspaces of H by $(S_h)_{h\in(0,1]}$, where we assume that S_h consists of spatially regular functions. The parameter $h \in (0, 1]$ governs the granularity of the spatial approximation, that is we expect to obtain a better approximation of elements in H if h gets smaller. In our example with $H = L^2([0, 1]; \mathbb{R})$, S_h may be a standard finite element space or the linear span of finitely many eigenfunctions of A (see Examples 3.6 and 3.7).

The spatially semidiscrete approximation $X_h : [0, T] \times \Omega \to S_h$ of the mild solution to (1.1) is given by the stochastic evolution equation

$$dX_h(t) + [A_h X_h(t) + P_h f(t, X_h(t))] \, dt = P_h g(t, X_h(t)) \, dW(t),$$

$$\text{for } 0 \leq t \leq T, \tag{1.3}$$

$$X_h(0) = P_h X_0,$$

where P_h denotes the orthogonal projector onto S_h and $A_h : S_h \to S_h$ is a discrete version of the operator A which will be defined in Sect. 3.2.

Next, we also introduce a spatio-temporal discretization of the stochastic evolution equation (1.1). Let $k \in (0, 1]$ denote a fixed time step which defines a time grid $t_j = jk$, $j = 0, 1, \ldots, N_k$, with $N_k k \leq T < (N_k + 1)k$.

Further, by X_h^j we denote the approximation of the mild solution X at time t_j. We combine the Galerkin methods with a linearly implicit Euler–Maruyama scheme and obtain the recursion, for $j = 1, \ldots, N_k$,

$$X_h^j - X_h^{j-1} + k\big(A_h X_h^j + P_h f(t_{j-1}, X_h^{j-1})\big) = P_h g(t_{j-1}, X_h^{j-1}) \Delta W^j,$$
$$X_h^0 = P_h X_0, \tag{1.4}$$

with the Wiener increments $\Delta W^j := W(t_j) - W(t_{j-1})$ which are $\mathscr{F}_{t_j}$-adapted, H-valued random variables. Consequently, X_h^j is itself an $\mathscr{F}_{t_j}$-adapted random variable which takes values in S_h.

Coming back to our computational task, it is therefore important to analyze the error

$$\big|\mathbf{E}\big[\varphi(X(T))\big] - \mathbf{E}\big[\varphi(X_h^{N_k})\big]\big|, \tag{1.5}$$

where the real-valued function $\varphi\colon H \to \mathbb{R}$ varies over a sufficiently large set of smooth test functions.

The error in (1.5) is called *weak error* and we say that the numerical approximation $X_h^{N_k}$ converges weakly to $X(T)$ for $h, k \to 0$, if the weak error vanishes for sufficiently many φ. In particular, if the weak error is supposed to be small, this forces the distribution of $X_h^{N_k}$ to be close to the distribution of $X(T)$.

While the approximation of the real number in (1.2) clearly corresponds to the weak convergence of the approximation method, the *strong error* plays an important role as well in this book. One reason for this is that it holds true by the mean-value theorem for Fréchet differentiable mappings that

$$\big|\mathbf{E}\big[\varphi(X(T)) - \varphi(X_h^{N_k})\big]\big|$$
$$= \left|\int_0^1 \mathbf{E}\big[\big(\varphi'\big(X_h^{N_k} + s(X(T) - X_h^{N_k})\big), X(T) - X_h^{N_k}\big)\big]\,\mathrm{d}s\right| \tag{1.6}$$
$$\leq C\big(1 + \|X(T)\|_{L^p(\Omega;H)}^{p-1} + \|X_h^{N_k}\|_{L^p(\Omega;H)}^{p-1}\big)\big\|X(T) - X_h^{N_k}\big\|_{L^p(\Omega;H)},$$

where we applied Hölder's inequality under the assumption that φ' satisfies a polynomial growth condition of the form $\|\varphi'(x)\| \leq C(1 + \|x\|^{p-1})$. Here, the so called *strong error*

$$\big\|X(T) - X_h^{N_k}\big\|_{L^p(\Omega;H)} \tag{1.7}$$

is measured with respect to the norm in $L^p(\Omega;H)$. Altogether, this shows that every strongly convergent approximation is also weakly convergent and the order of convergence of the strong error is always a lower bound for the order of the weak error.

Further, a strongly convergent scheme ensures that important features of single sample paths of the solution are reproduced by its approximation. In particular, strong convergence also implies the pathwise convergence analyzed in [15, 42, 43].

Finally, as it was shown by Giles [27, 28], the strong order of convergence is also essential for developing efficient multilevel Monte Carlo methods for applications where the weak approximation is of interest in the first place. SPDE related references are [4, 5].

The following three sections give a more detailed overview of the results covered in this monograph. The first is concerned with the spatial and temporal regularity of the mild solution X to (1.1) and presents the main idea behind the technique used in Sects. 2.5 and 2.6. In the remaining two sections we sketch our strong and weak convergence results.

1.1 Optimal Regularity Results

Before we introduce our discretization schemes it is important to develop a deeper understanding of the properties of X itself. We will do this in all details in the Sects. 2.3–2.6. Of particular importance for the error analysis of numerical approximations is the spatial and temporal regularity of the mild solution, since this constitutes a natural bound on the order of convergence. Therefore, in order to obtain an optimal result on the order of convergence it is crucial to have a precise knowledge of the regularity.

The spatial and temporal regularity of the mild solution to (1.1) has already been extensively studied in the literature. For example, we refer to [18] for results on the stochastic convolution. Often, these results are derived under the assumption that $-A$ in (1.1) is the generator of an analytic semigroup and that the nonlinearity g satisfies a global Lipschitz assumption on H. In this situation it is shown that the mild solution takes values in $\dot{H}^{1-\epsilon}$ for every $\epsilon \in (0, 1]$, where $\dot{H}^s :=$ $\mathrm{dom}(A^{\frac{s}{2}})$, $s \geq 0$, denotes the domain of fractional powers of the operator A (see Appendix B.2).

Under our additional assumption, that A is self-adjoint, positive definite and with compact inverse, it is also well-known, for example in [17, 18, Chap. 6.3] and [67], that the ϵ can be removed from the spatial regularity result and that the mild solution takes indeed values in $\dot{H}^1 = \mathrm{dom}(A^{\frac{1}{2}})$ with probability one.

On the other hand, in the recent paper [41] the authors work in the more general situation, where $-A$ is only assumed to be the generator of an analytic semigroup. But, in addition to the global Lipschitz condition, they impose a linear growth bound on the nonlinearity $g \colon H \to L(H)$ of the form $\|A^{\frac{r}{2}}g(x)\|_{L_2^0} \leq C(1 + \|A^{\frac{r}{2}}x\|)$ for some $r \in [0, 1)$ and all $x \in \dot{H}^r = \mathrm{dom}(A^{\frac{r}{2}})$, where $\|\cdot\|_{L_2^0}$ denotes the Hilbert–Schmidt operator norm (see (2.10) and Assumption 2.20). In this case, the authors show that the mild solution takes values in $\dot{H}^{1+r-\epsilon}$ for every $\epsilon \in (0, 1 + r]$ with probability one.

While a corresponding global Lipschitz condition on g as a mapping on $\dot{H}^r$ may force g to be affine linear, the linear growth bound is also satisfied by several Nemytskii operators as it is shown in [41, Sect. 4]. Therefore, the linear growth bound has proven itself to be useful in order to obtain sharper estimates of the spatial regularity for semilinear stochastic evolution equations.

In our regularity analysis in Sects. 2.5 and 2.6 we work under both additional assumptions, that is we study the linear growth bound from [41] under the assumption that A is self-adjoint and positive definite with compact inverse. As our first truly new result, we show in Sects. 2.5 and 2.6 that the ϵ can also be removed in this situation. Hence, $X(t)$ takes values in $\dot{H}^{1+r}$ with probability one (see Theorem 2.27).

The two main ingredients of our regularity analysis are sharp integral versions of the smoothing property of the semigroup (see Lemma B.9(*iii*)) which are explicitly tailored to estimate the regularity of the convolution. Secondly, we employ the following technique in order to estimate the norms of stochastic convolutions. Let $p \geq 2$. After applying a Burkholder–Davis–Gundy-type inequality (see Proposition 2.12) we obtain

$$\left\| A^{\frac{1+r}{2}} \int_0^t E(t-\sigma)g(\sigma, X(\sigma))\, dW(\sigma) \right\|_{L^p(\Omega;H)}$$
$$\leq C \left(\mathbf{E}\left[\left(\int_0^t \left\| A^{\frac{1+r}{2}} E(t-\sigma)g(\sigma, X(\sigma)) \right\|_{L_2^0}^2 d\sigma \right)^{\frac{p}{2}} \right] \right)^{\frac{1}{p}}.$$

At this point Lemma B.9(*iii*) is not directly applicable since $g(\sigma, X(\sigma))$ is obviously not independent of σ. However, since

$$\left(\mathbf{E}\left[\left(\int_0^t \left\| A^{\frac{1+r}{2}} E(t-\sigma)g(\sigma, X(\sigma)) \right\|_{L_2^0}^2 d\sigma \right)^{\frac{p}{2}} \right] \right)^{\frac{1}{p}}$$
$$\leq \left(\mathbf{E}\left[\left(\int_0^t \left\| A^{\frac{1+r}{2}} E(t-\sigma)\big(g(\sigma, X(\sigma)) - g(t, X(t))\big) \right\|_{L_2^0}^2 d\sigma \right)^{\frac{p}{2}} \right] \right)^{\frac{1}{p}}$$
$$+ \left(\mathbf{E}\left[\left(\int_0^t \left\| A^{\frac{1+r}{2}} E(t-\sigma)g(t, X(t)) \right\|_{L_2^0}^2 d\sigma \right)^{\frac{p}{2}} \right] \right)^{\frac{1}{p}}$$

Lemma B.9(*iii*) is applicable to the second summand and the first can by estimated by Lemma B.9(*i*) together with the Hölder continuity of the mapping $\sigma \mapsto g(\sigma, X(\sigma))$. For details we refer to the proof of Lemma 2.29.

Our technique is already known in the literature for Bochner integrals consisting of a convolution with an analytic semigroup, for example in [19, Prop. 3] and [65, p. 157]. However, to the best of our knowledge, this technique appeared for the first time in [50] in order to derive regularity estimates of stochastic convolutions.

In this book, we slightly extend the results of [50] to nonlinearities f and g which are allowed to depend on $t \in [0, T]$ and $\omega \in \Omega$.

1.2 Optimal Strong Error Estimates

After we have established the regularity estimates, our second main result is concerned with the optimal order of convergence of the strong error for the spatially semidiscrete approximation (1.3) and the spatio-temporal discretization

(1.4) of (1.1). For the last 15 years this has been a very active field of research and for an extensive list of references we refer to the review article [38].

We basically work under the same assumptions as for the results on optimal regularity in Sects. 2.5 and 2.6. That is, there exists a parameter $r \in [0, 1)$ which, in particular, controls the spatial regularity of the mild solution X such that $X(t) \in \dot{H}^{1+r}$ with probability one for all $t \in [0, T]$.

In this situation, we prove in Sect. 3.4 that the spatially semidiscrete approximation (1.3) is strongly convergent and for all $p \geq 2$ it holds for a constant C, which is independent of h, that

$$\|X_h(t) - X(t)\|_{L^p(\Omega;H)} \leq C h^{1+r}, \text{ for all } t \in (0, T].$$

Therefore, in our example of a standard finite element semidiscretization, the approximation X_h converges with order $1 + r$ to the mild solution X. Since this rate coincides with the spatial regularity of X it is called optimal (see [66, Chap. 1]).

We stress that, to the best of our knowledge, in all articles, which deal with the numerical approximation of semilinear stochastic partial differential equations, the obtained order of convergence is of the suboptimal form $1 + r - \epsilon$ for any $\epsilon > 0$ (see [70] or [34], where also stronger Lipschitz assumptions have been imposed on f, g) or the error estimates contain a logarithmic term of the form $\log(t/h)$ as in [44].

A similar result (see Theorem 3.14) also holds for the spatio-temporal discretization (1.4) of X, namely, there exists a constant C, independent of $k, h \in (0, 1]$, such that

$$\|X_h^j - X(t_j)\|_{L^p(\Omega;H)} \leq C(h^{1+r} + k^{\frac{1}{2}}), \quad \text{for all } j = 1, \ldots, N_k.$$

Again, we obtain the optimal order of convergence with respect to the spatio-temporal discretization.

It is a well-known fact [13], that any one-step method, which only uses the information of the driving Wiener process provided by the increments $\Delta W^j = W(t_j) - W(t_{j-1})$, has the maximum strong order of convergence $\frac{1}{2}$ with respect to the time step k. It is possible to overcome this barrier by considering Itô–Taylor approximations as in [39]. In particular, Milstein-like schemes are discussed in the recent papers [3, 40, 48, 51].

In principle, these optimal error estimates are derived by a similar technique as sketched above for the optimal regularity results. Indeed, in Sects. 3.3 and 3.5 we present several lemmas which play the same crucial role as Lemma B.9(*iii*) and (*iv*) in the proof of the optimal spatial regularity. All these lemmas are concerned with the spatial semidiscretization and the fully discrete approximation of the deterministic homogeneous equation

$$\frac{d}{dt}u(t) + Au(t) = 0, \quad t > 0, \quad \text{with } u(0) = x. \tag{1.8}$$

In particular, these lemmas are tailored to estimate the integral of the square-norm of the discretization error, as it appears in estimates of stochastic integrals. For the fully discrete approximation these lemmas are to some extend new and have first appeared in [49].

1.3   A New Approach to Weak Convergence

In the context of SPDEs, the study of the weak error of convergence has gained broad attention only recently and the literature is much less extensive compared to the analysis of the strong error. Without giving a complete list, important contributions were achieved in [21, 22, 26, 35, 45, 46] and the references therein.

So far, the usual approach in the literature to analyze the weak error involves the study of the solution to the *Kolmogorov equation* associated to the stochastic evolution equation. Unlike the finite dimensional case, which leads to a deterministic partial differential equation on a finite dimensional spatial domain, the Kolmogorov equation associated to a stochastic evolution equation on a Hilbert space H is in fact a PDE on the infinite dimensional space H, see for example [63, 64] and the references therein.

In particular for semilinear stochastic evolution equations, the analysis of its Kolmogorov equation turns out to be involved. Since the derivation of growth bounds for various derivatives of the solution to the Kolmogorov equation is crucial in the weak error analysis, one usually needs to invoke assumptions on f and g which are much more restrictive compared to the analysis of the strong error.

For example, in [21] the author assumes that $f: H \rightarrow H, g: H \rightarrow L(H)$ are deterministic, time independent and three times continuously Fréchet differentiable with bounded derivatives. Under a further, more technically motivated assumption on g, namely that

$$\|g''(x)[y, y]\|_{L(H)} \leq C \|A^{-\frac{1}{4}} y\|^2 \quad \text{for all } x, y \in H,$$

the author proves that a temporal semidiscrete approximation based on the implicit Euler scheme converges weakly with order $\frac{1}{2} - \epsilon, \epsilon > 0$ arbitrary small, to the mild solution of (1.1). For this, the Wiener process is assumed to be a cylindrical white noise with $Q = \mathrm{Id}_H$ and, besides the Kolmogorov equation, the author also applies techniques from the Malliavin calculus in order to obtain this result in [21].

Note that in the same situation, but under much less severe regularity assumptions on f and g, the strong error is known to converge with order $\frac{1}{4} - \epsilon$, for every $\epsilon > 0$ arbitrary small.

In these notes we propose a new approach to analyze the weak error of convergence, which solely builds on methods from the Malliavin calculus and completely avoids the solution to the Kolmogorov equation. By these means we hope to obtain weak convergence results with optimal order in much more general situations, in particular if f and g are also allowed to depend on $\omega \in \Omega$.

However, so far we are only able to apply our ansatz to linear stochastic evolution equations with additive noise, that is

$$dX(t) + \big[AX(t) + f(t)\big]\,dt = g(t)\,dW(t), \quad \text{for } 0 \le t \le T,$$
$$X(0) = X_0. \tag{1.9}$$

This equation is understood in the same sense as (1.1), but now the inhomogeneities $f : [0, T] \times \Omega \to H$ and $g : [0, T] \times \Omega \to L(H)$ are stochastic processes which do not directly depend on $X(t)$. Our more detailed assumptions are given in Sect. 5.2.

The starting point of our ansatz is again (1.6), which this time is considered for the semidiscrete approximation $X_h(T)$. Just before we applied the Hölder inequality in (1.6) we have

$$\big|\mathbf{E}\big[\varphi(X_h(T)) - \varphi(X(T))\big]\big|$$
$$= \bigg| \int_0^1 \mathbf{E}\big[\big(\varphi'(X(T) + s(X_h(T) - X(T))), X_h(T) - X(T)\big)\big]\,ds \bigg|. \tag{1.10}$$

Here we insert the variation of constants formula for $X(T)$ and $X_h(T)$, which read

$$X(T) = E(T)X_0 - \int_0^T E(T - \sigma)f(\sigma)\,d\sigma + \int_0^T E(T - \sigma)g(\sigma)\,dW(\sigma)$$

and

$$X_h(T) = E_h(T)X_0 - \int_0^T E_h(T - \sigma)P_h f(\sigma)\,d\sigma + \int_0^T E_h(T - \sigma)P_h g(\sigma)\,dW(\sigma),$$

where $(E(t))_{t \in [0,T]}$ and $(E_h(t))_{t \in [0,T]}$ denote the strongly continuous semigroups generated by $-A$ and $-A_h$, respectively. Then, we get

$$\mathbf{E}\big[\varphi(X_h(T)) - \varphi(X(T))\big]$$
$$= \int_0^1 \mathbf{E}\big[\big(\varphi'(X(T) + s(X_h(T) - X(T))), X_h(T) - X(T)\big)\big]\,ds$$
$$= \int_0^1 \mathbf{E}\bigg[\bigg(\varphi'(X(T) + s(X_h(T) - X(T))), F_h(T)X_0$$
$$- \int_0^T F_h(T - \sigma)f(\sigma)\,d\sigma\bigg)\bigg]\,ds$$
$$+ \int_0^1 \mathbf{E}\bigg[\bigg(\varphi'(X(T) + s(X_h(T) - X(T))), \int_0^T F_h(T - \sigma)g(\sigma)\,dW(\sigma)\bigg)\bigg]\,ds,$$

where $F_h(t) := E_h(t)P_h - E(t)$ denotes the error operator of the spatial discretization. If we apply Hölder's inequality to the first summand, we obtain the optimal order of weak convergence. However, the estimate of the second summand is more delicate and the Malliavin calculus comes into play. The idea is to apply Theorem 4.13, the so called *Bismut's integration by parts formula*, which yields

$$\mathbf{E}\Big[\Big(F, \int_0^T \Psi(t)\,\mathrm{d}W(t)\Big)\Big] = \mathbf{E}\big[(DF, \Psi)_{L^2([0,T];L_2^0)}\big]. \tag{1.11}$$

Here, $F: \Omega \to H$ is smooth in the sense of the Malliavin calculus and D denotes the Malliavin derivative. In Chap. 4 we review the most important properties of D and present a proof of (1.11).

We follow this idea in Sect. 5.3 and derive a representation formula of the weak error (see Theorems 5.9 and 5.10) which is then used to prove the weak order of convergence for the spatially semidiscrete approximation and the spatio-temporal discretization of $X(T)$. We validate the well-known rule of thumb for the weak approximation of SODEs, that under quite general assumptions on f and g the order of convergence of the weak error is almost twice the order of the strong error.

Let us remark that a similar error representation formula is given in [45], which in contrast to our formula contains derivatives of the solution to the Kolmogorov equation. However, both formulas yield the same error estimates if applied to linear stochastic partial differential equations such as the stochastic heat equation, the stochastic wave equation or the linearized stochastic Cahn–Hilliard equation as it is shown in [45].

1.4   Outline

In this section we give a short overview of the remaining parts of this book.

Chapter 2 is mainly concerned with the study of the stochastic evolution equation (1.1). In the first two sections we recall some useful facts on Wiener processes and stochastic Itô-integrals which take values in separable Hilbert spaces. The content of these sections is primarily based on [18, Chaps. 3 and 4] and [61, Chap. 2]. In Sect. 2.3 we describe all assumptions and details which are concerned with the semilinear stochastic evolution equation (1.1). This section also includes several examples.

In the remainder of this chapter we prove an existence and uniqueness result in Sect. 2.4, where our proof is a slightly generalized and simplified version of the proof of [41, Th. 1]. In Sects. 2.5 and 2.6 we present our optimal spatial and temporal regularity results. The final section of Chap. 2 contains some indications to possible generalizations of our results.

Chapter 3 is devoted to the analysis of the strong error. After some preliminaries in the first section, we provide a brief introduction into the Galerkin finite element methods. In particular, we introduce the discrete analogue of A and provide two more concrete examples of spatial discretizations.

Sections 3.3 and 3.4 are concerned with the spatially semidiscrete approximation of the stochastic evolution equation. First, we derive sharp integral versions of well-known convergence results from [66] and prove some extensions to non-smooth initial data. Then, we give the proof of our result on the optimal order of strong convergence for the spatial semidiscretization. The Sects. 3.5 and 3.6 follow the same path but are concerned with the strong convergence of the spatio-temporal discretization.

In Chap. 4 we give an overview of the fundamentals of the Malliavin Calculus for Hilbert space valued stochastic processes. This section is based on [32, 54]. As a nice application of Bismut's integration by parts formula, which is presented in Sect. 4.2, we give a short proof of the well-known stochastic Fubini theorem [18, Th. 4.18] in Sect. 4.3.

Chapter 5 deals with the weak order of convergence of the spatially semidiscrete approximation and spatio-temporal discretization of the linear stochastic evolution equation (1.9). After some preliminaries and the precise formulation of the assumptions, we first derive for both approximation methods a representation formula of the weak error in Sect. 5.3. In Sect. 5.4 we apply this formula in order to analyze the weak error for the inhomogeneous heat equation.

The final Chap. 6 contains four numerical experiments, where we illustrate the validity of our theoretical findings for the spatially semidiscrete approximation. In particular, our experiments consider the stochastic heat equation with and without inhomogeneities in Sects. 6.2 and 6.3. Further, in Sects. 6.4 and 6.5 we are concerned with the geometric Brownian motion in infinite dimensions, again with and without inhomogeneities.

The appendix contains some auxiliary results, where Appendix A is concerned with several variations of the classical Gronwall-Lemma. In Appendix B we collect several results on semigroups with unbounded infinitesimal generator. The perhaps most important results are the integral versions of the smoothing property in Lemma B.9, which have many consequences throughout this monograph. In the final appendix, we cite a generalized version of Lebesgue's dominated convergence theorem from [2, 1.23].

Chapter 2
Stochastic Evolution Equations in Hilbert Spaces

This chapter deals with semilinear stochastic evolution equations in Hilbert spaces. The first two sections are more introductory and review some important properties of Wiener processes and the stochastic Itô integral in Hilbert spaces. The main references for the presented material are [18, 61].

In Sect. 2.3 we introduce the general form of the semilinear stochastic evolution equations from [18, Chap. 7.1] which we treat numerically in the following chapters. Besides some more explicit examples this section also contains our usual set of assumptions which we use in order to derive all results in this chapter.

In Sect. 2.4 we present an existence and uniqueness result. Although our assumptions on the nonlinearities are slightly more general as in [41] the presented proof follows basically the same idea as the proof of [41, Th. 1]. But since we give a more detailed regularity analysis of the mild solution in the two subsequent sections, our proof is slightly simplified.

In Sects. 2.5 and 2.6 we precisely determine the spatial and temporal regularity properties of the mild solution. These results first appeared in [50].

In the final section of this chapter, we also discuss some possible generalizations of our results.

2.1 Hilbert Space-Valued Wiener Processes

This section gives a short review on Hilbert space-valued Wiener processes. All results are well-known in the literature, for example, in [18, Chap. 4.1] and [61, Chap. 2.1].

By $(\Omega, \mathscr{F}, \mathbf{P})$ we denote a probability space and by $(U, (\cdot, \cdot)_U, \|\cdot\|_U)$ a separable Hilbert space. We also consider a linear, bounded, self-adjoint, positive semidefinite operator $Q \in L(U)$. The trace of Q is given by

$$\mathrm{Tr}(Q) := \sum_{i \in \mathbb{N}} (Qe_i, e_i)_U, \qquad (2.1)$$

R. Kruse, *Strong and Weak Approximation of Semilinear Stochastic Evolution Equations*,
Lecture Notes in Mathematics 2093, DOI 10.1007/978-3-319-02231-4_2,
© Springer International Publishing Switzerland 2014

where $(e_i)_{i \in \mathbb{N}}$ is an arbitrary orthonormal basis of U. If the series $\mathrm{Tr}(Q)$ is (absolutely) convergent, the operator Q is called *trace class* and the real number $\mathrm{Tr}(Q) < \infty$ does not depend on the particular choice of the orthonormal basis.

Further, from [18, Prop. C.3] it follows that every self-adjoint, positive semidefinite operator $Q \in L(U)$ with finite trace is compact and, therefore, the spectral theorem for compact operators [69, Th. VI.3.2] yields the existence of an orthonormal basis $(e_i)_{i \in \mathbb{N}}$ of U and a decreasing sequence of nonnegative real numbers $(\mu_i)_{i \in \mathbb{N}}$ with $\mu_i \to 0$ as $i \to \infty$ such that

$$Q e_i = \mu_i e_i, \quad \text{for all } i \in \mathbb{N},$$

and

$$Q u = \sum_{i \in \mathbb{N}} \mu_i (u, e_i)_U e_i, \quad \text{for all } u \in U.$$

In particular, it holds

$$\mathrm{Tr}(Q) = \sum_{i \in \mathbb{N}} \mu_i.$$

We now define a (standard) Q-Wiener process in the same way as in [18, Chap. 4.1] and [61, Def. 2.1.9].

Definition 2.1. Let $T > 0$. A stochastic process $W \colon [0, T] \times \Omega \to U$ on $(\Omega, \mathscr{F}, \mathbf{P})$ is called a *(standard) Q-Wiener process* if

(i) $W(0) = 0$,
(ii) W has **P**-a.s. continuous trajectories,
(iii) W has independent increments, that is, for all $n \in \mathbb{N}$ and all partitions $0 \leq t_1 < \cdots < t_n \leq T$ it holds that the random variables

$$W(t_1), W(t_2) - W(t_1), \ldots, W(t_n) - W(t_{n-1})$$

are independent,
(iv) for all $0 \leq s < t \leq T$ the increment $W(t) - W(s)$ is a Gaussian random variable with mean $0 \in U$ and *covariance operator* $(t - s)Q$, that is

$$\mathbf{P} \circ \big(W(t) - W(s)\big)^{-1} = N(0, (t - s)Q).$$

For the definition of the Gaussian law $N(0, (t - s)Q)$ on Hilbert spaces we refer to [18, Chap. 2.3.2] and [61, Def. 2.1.1]. Here, we just recall the following important properties of a U-valued Gaussian random variable from [61, Prop. 2.1.4]:

Proposition 2.2 ([61, Prop. 2.1.4]). *Consider a U-valued Gaussian random variable X with mean $m \in U$ and covariance operator $Q \in L(U)$, where Q is self-adjoint, positive semidefinite and with finite trace, that is $\mathbf{P} \circ X^{-1} = N(m, Q)$.*

Then, for all $u \in U$, $(X, u)_U$ is a real-valued Gaussian random variable with

(i) $\mathbf{E}\big[(X, u)_U\big] = (m, u)_U$ *for all $u \in U$,*
(ii) $\mathbf{E}\big[(X - m, u)_U (X - m, v)_U\big] = (Qu, v)_U$ *for all $u, v \in U$,*
(iii) $\mathbf{E}\big[\|X - m\|_U^2\big] = \mathrm{Tr}(Q)$.

The following representation of a Q-Wiener process turns out to be very useful.

Proposition 2.3 ([61, Prop. 2.1.10]). *Consider a self-adjoint, positive semidefinite operator $Q \in L(U)$. Let $(e_i)_{i \in \mathbb{N}}$ denote an orthonormal basis of U consisting of eigenvectors of Q with corresponding eigenvalues $(\mu_i)_{i \in \mathbb{N}}$. Then a stochastic process $W : [0, T] \times \Omega \to U$ is a Q-Wiener process if and only if*

$$W(t) = \sum_{i \in \mathbb{N}} \sqrt{\mu_i} \beta_i(t) e_i, \quad \text{for all } t \in [0, T], \tag{2.2}$$

where β_i, for all i with $\mu_i > 0$, are independent real-valued Brownian motions on $(\Omega, \mathscr{F}, \mathbf{P})$.

The series (2.2) also converges in $L^2(\Omega; C([0, T]; U))$ and thus always has a $\mathbf{P}$-a.s. continuous modification. In particular, under the above conditions on Q there always exists a Q-Wiener process on U.

The proof is given in [61, Prop. 2.1.10]. The next definition follows the lines of [61, Def. 2.1.12].

Definition 2.4. A Q-Wiener process $W : [0, T] \times \Omega \to U$ is called a *Q-Wiener process with respect to a filtration* $(\mathscr{F}_t)_{t \in [0,T]}$, if

(i) W is adapted to $(\mathscr{F}_t)_{t \in [0,T]}$,
(ii) $W(t) - W(s)$ is independent of $\mathscr{F}_s$ for all $0 \leq s < t \leq T$.

Any Q-Wiener process $W : [0, T] \times \Omega \to U$ is in fact a Q-Wiener process with respect to a *normal filtration*, as it is shown in [61, Prop. 2.1.13]. Here a normal filtration is a filtration $(\mathscr{F}_t)_{t \in [0,T]}$ such that $\mathscr{F}_0$ contains all sets $A \in \mathscr{F}$ with $\mathbf{P}(A) = 0$ and

$$\mathscr{F}_t = \mathscr{F}_{t+} = \bigcap_{s > t} \mathscr{F}_s, \quad \text{for all } t \in [0, T].$$

The next proposition from [61, Prop. 2.2.10] is concerned with the martingale property of Wiener processes. As in [61, Chap. 2.2] a stochastic process $M : [0, T] \times \Omega \to B$, where B denotes a separable Banach space, is called an $(\mathscr{F}_t)_{t \in [0,T]}$-*martingale* if M is adapted with $\mathbf{E}\big[\|M(t)\|_B\big] < \infty$ for all $t \in [0, T]$ and

$$\mathbf{E}\big[M(t)|\mathscr{F}_s\big] = M(s) \quad \mathbf{P}\text{-a.s. for all } 0 \leq s \leq t \leq T,$$

where as usual [61, Prop. 2.2.1], $\mathbf{E}[\cdot|\mathscr{G}]$ denotes the *conditional expectation* with respect to a σ-field $\mathscr{G} \subset \mathscr{F}$.

Proposition 2.5 ([61, Prop. 2.2.10]). *Let* $W : [0, T] : \Omega \rightarrow U$ *be a* Q-*Wiener process with respect to a normal filtration* $(\mathscr{F}_t)_{t \in [0,T]}$ *on* $(\Omega, \mathscr{F}, \mathbf{P})$. *Then* W *is a continuous square-integrable* $(\mathscr{F}_t)_{t \in [0,T]}$-*martingale.*

For the proof and further results on martingales in Banach spaces we refer to [61, Chap. 2.2].

So far, all results rely on the assumption that the covariance operator $Q \in L(U)$ of the Wiener process has finite trace. In order to also deal with the *white noise* case, that is $Q = \mathrm{Id}_U$, this assumption turns out to be too restrictive. However, as it is shown in [18, Chap. 4.3.1] and [61, Chap. 2.5.1] it is possible to define Q-Wiener processes for more general covariance operators $Q \in L(U)$.

For this we first cite the following result from [61, Prop. 2.3.4] which in turn is based on [62, Th. VI.9]. For a proof we also refer to [69, Bsp. (d), Kor. VII.1.16].

Proposition 2.6 ([61, Prop. 2.3.4]). *Let* $Q \in L(U)$ *be positive semidefinite and self-adjoint. Then there exists a unique self-adjoint and positive semidefinite operator* $Q^{\frac{1}{2}} \in L(U)$ *such that* $Q^{\frac{1}{2}} \circ Q^{\frac{1}{2}} = Q$.

Then, as in [18, Chap. 4.2] and [61, Chap. 2.3], it is useful to introduce the space $U_0 := Q^{\frac{1}{2}}(U)$ which together with the inner product

$$(u_0, v_0)_{U_0} := \left(Q^{-\frac{1}{2}} u_0, Q^{-\frac{1}{2}} v_0 \right)_U, \quad \text{for all } u_0, v_0 \in U_0, \tag{2.3}$$

becomes a separable Hilbert space. If Q is not one-to-one $Q^{-\frac{1}{2}}$ denotes the pseudo inverse of $Q^{\frac{1}{2}}$ [61, App. C].

Next, by following the lines of [61, Chap. 2.5.1], let us introduce a further Hilbert space $(U_1, (\cdot, \cdot)_{U_1}, \| \cdot \|_{U_1})$ and a Hilbert–Schmidt embedding $J : U_0 \to U_1$, that is

$$\sum_{i=1}^{\infty} \| J e_i \|_{U_1}^2 < \infty$$

for one (and consequently all) orthonormal basis $(e_i)_{i \in \mathbb{N}}$ of U_0.

As it was pointed out in [61, Rem. 2.5.1] such a Hilbert space U_1 and the embedding J always exists. The next proposition states the existence of a so called *cylindrical* Q-*Wiener process* in U.

Proposition 2.7 ([61, Prop. 2.5.2]). *Let* $(e_i)_{i \in \mathbb{N}}$ *denote an orthonormal basis of* $U_0 = Q^{\frac{1}{2}}(U)$ *and* $(\beta_i)_{i \in \mathbb{N}}$ *a sequence of independent, real-valued Brownian motions. If we set* $Q_1 := JJ^*$ *it holds that* $Q_1 \in L(U_1)$ *is positive semidefinite, self-adjoint and of finite trace and it holds* $Q_1^{\frac{1}{2}}(U_1) = J(U_0)$. *In addition, we have*

$$\| u_0 \|_{U_0} = \left\| Q_1^{-\frac{1}{2}} J u_0 \right\|_{U_1} = \| J u_0 \|_{Q_1^{\frac{1}{2}}(U_1)}. \tag{2.4}$$

Hence, $J : U_0 \to Q_1^{\frac{1}{2}}(U_1)$ *is an isometry, where the norm* $\| \cdot \|_{Q_1^{\frac{1}{2}}(U_1)}$ *is defined in the same way as in* (2.3).

Further, the stochastic process $W : [0, T] \times \Omega \to U_1$ given by

$$W(t) = \sum_{i=1}^{\infty} \beta_i(t) J e_i, \quad \text{for all } t \in [0, T],$$ (2.5)

is a standard Q_1-Wiener process on U_1. The series in (2.5) even converges with respect to the norm in $L^2(\Omega; C([0, T]; U_1))$.

2.2 The Hilbert Space-Valued Stochastic Itô-Integral

In this section we collect some important properties of the stochastic Itô-integral. Let $(\Omega, \mathscr{F}, \mathbf{P})$ be a probability space and $T > 0$. We consider a standard Q-Wiener process $W : [0, T] \times \Omega \to U$ with respect to a normal filtration $(\mathscr{F}_t)_{t \in [0,T]}$, where $(U, (\cdot, \cdot)_U, \| \cdot \|_U)$ denotes a separable Hilbert space and the covariance operator $Q \in L(U)$ is positive-semidefinite and self-adjoint. For the first part of this section we also assume that Q is of finite trace.

Let $(H, (\cdot, \cdot), \| \cdot \|)$ denote a further Hilbert space. Then the H-valued *stochastic Itô-integral* of a stochastic process $\Phi : [0, T] \times \Omega \to L(U, H)$ with respect to the Wiener process W is denoted by

$$\int_0^T \Phi(\sigma) \, dW(\sigma).$$

Without going into the details of the construction we only recall from [61, Chap. 2.3.2] that the integral is first defined in terms of *elementary integrands* of the form

$$\Phi(t) = \sum_{i=0}^{n-1} \Phi_i 1_{(t_i, t_{i+1}]}(t), \quad \text{for } t \in [0, T],$$ (2.6)

where $n \in \mathbb{N}$ and $\Phi_i : \Omega \to L(U, H)$ are $\mathscr{F}_{t_i}$-measurable with respect to the strong Borel-σ-field on $L(U, H)$ for $0 \le i \le n - 1$ and all Φ_i only take a finite number of values in $L(U, H)$. Then the stochastic integral of Φ is given by

$$\int_0^T \Phi(\sigma) \, dW(\sigma) := \sum_{i=0}^{n-1} \Phi_i \big(W(t_{i+1}) - W(t_i)\big).$$ (2.7)

In the next step one defines a norm on the set of all elementary integrands such that the Itô-integral becomes an isometry between these integrands and the space of all H-valued continuous square-integrable martingales with respect to $(\mathscr{F}_t)_{t \in [0,T]}$.

For this norm the notion of *Hilbert–Schmidt operators* plays an important role.

Definition 2.8 (Hilbert–Schmidt operator [61, Def. 2.3.3]). Let U and H denote separable Hilbert spaces. An operator $A \in L(U, H)$ is called *Hilbert–Schmidt* if

$$\sum_{i=1}^{\infty} \|Ae_i\|^2 < \infty,$$

where $(e_i)_{i \in \mathbb{N}}$ denotes an arbitrary orthonormal basis of U.

The set of all Hilbert–Schmidt operators $A \colon U \to H$ is denoted by $L_2(U, H)$. Together with the inner product

$$(A, B)_{L_2(U,H)} := \sum_{i=1}^{\infty} (Ae_i, Be_i), \quad \text{for } A, B \in L_2(U, H), \tag{2.8}$$

$L_2(U, H)$ becomes a separable Hilbert space. For short we write $L_2(U) = L_2(U, U)$.

By Werner [69, Satz VI.6.2] the inner product and the induced *Hilbert–Schmidt norm* in $L_2(U, H)$

$$\|A\|_{L_2(U,H)} := \Big(\sum_{i=1}^{\infty} \|Ae_i\|^2 \Big)^{\frac{1}{2}} \tag{2.9}$$

does not depend on the particular choice of the orthonormal basis $(e_i)_{i \in \mathbb{N}}$ in U.

As it is shown in [61, Rem. B.0.6] Hilbert–Schmidt operators enjoy the following properties. A proof is also given in [69, Satz VI.6.2].

Proposition 2.9. *Let $A \in L_2(U, H)$ be a Hilbert–Schmidt operator:*

 (i) *The adjoint A^* of A is also a Hilbert–Schmidt operator and it holds*

$$\|A\|_{L_2(U,H)} = \|A^*\|_{L_2(H,U)}.$$

 (ii) *It holds that $\|A\|_{L(U,H)} \le \|A\|_{L_2(U,H)}$.*

(iii) *Consider further separable Hilbert spaces G_1, G_2. For all $T_1 \in L(G_1, U)$ and $T_2 \in L(H, G_2)$ it holds $T_2 A T_1 \in L_2(G_1, G_2)$ with*

$$\|T_2 A T_1\|_{L_2(G_1,G_2)} \le \|T_2\|_{L(H,G_2)} \|A\|_{L_2(U,H)} \|T_1\|_{L(G_1,U)}.$$

Before we come to the above mentioned isometry for the stochastic Itô-integral, let us also introduce the notation $L_2^0 := L_2(U_0, H)$. If Q is trace class we have that

$$\|A\|_{L_2^0} = \|A \circ Q^{\frac{1}{2}}\|_{L_2(U,H)} \quad \text{for all } A \in L(U, H), \tag{2.10}$$

since it holds that $Q^{\frac{1}{2}} \in L_2(U)$ with $\|Q^{\frac{1}{2}}\|_{L_2(U)} = \mathrm{Tr}(Q)$ and $\{A|_{U_0} \mid A \in L(U, H)\} \subset L_2^0$ by Proposition 2.9(iii).

The so called *Itô-isometry* [61, Prop. 2.3.5] is now given by

$$\mathbf{E}\left[\left\|\int_0^T \Phi(\sigma)\,\mathrm{d}W(\sigma)\right\|^2\right] = \mathbf{E}\left[\int_0^T \|\Phi(\sigma)\|_{L_2^0}^2\,\mathrm{d}\sigma\right] =: \|\Phi\|_T^2 \qquad (2.11)$$

which holds for all elementary integrands Φ of the form (2.7).

Next, the set $\mathcal{N}_W^2(0,T;H)$ of all integrable stochastic processes is given as the abstract completion of all elementary integrands of the form (2.7) in $L^2([0,T] \times \Omega, \mathcal{B}([0,T]) \otimes \mathcal{F}, \mathrm{d}t \otimes \mathbf{P}; L(U,H))$ with respect to the norm $\|\cdot\|_T$.

In fact, this extension of the stochastic Itô-integral remains isometric and is unique [61, Chap. 2.3, p. 27]. Further, in [61, Chap. 2.3, p. 28] the characterization

$$\mathcal{N}_W^2(0,T;H) = L^2([0,T] \times \Omega, \mathscr{P}_T, \mathrm{d}t \otimes \mathbf{P}; L_2^0) \qquad (2.12)$$

is given, where $\mathscr{P}_T$ denotes the σ-field of *predictable stochastic processes*, that is

$$\mathscr{P}_T := \sigma\left(\{(s,t] \times F_s \mid 0 \le s < t \le T,\ F_s \in \mathcal{F}_s\} \cap \{\{0\} \times F_0 \mid F_0 \in \mathcal{F}_0\}\right)$$

$$= \sigma\left(Y : [0,T] \times \Omega \to \mathbb{R} \mid Y \text{ is left-continuous and adapted to } \mathcal{F}_t,\ t \in [0,T]\right).$$

Let us remark that by the so-called *localization procedure* one can further extend the domain of the stochastic Itô-integral to an even larger set of stochastic processes. However, in the following this generalization plays no important role and for all details we refer to [61, Chap. 2.3].

The following properties of the stochastic Itô-integral will be useful.

Proposition 2.10 ([18, Th. 4.12, Prop. 4.13, Cor. 4.14]). *For all $\Phi \in \mathcal{N}_W^2(0,T;H)$ the stochastic process*

$$M(t) := \int_0^t \Phi(\sigma)\,\mathrm{d}W(\sigma), \quad t \in [0,T],$$

is a continuous, square integrable martingale with

$$\mathbf{E}[M(t)] = 0, \quad \mathbf{E}[\|M(t)\|^2] = \int_0^t \|\Phi(\sigma)\|_{L_2^0}^2\,\mathrm{d}\sigma$$

and for $\Phi_1, \Phi_2 \in \mathcal{N}_W^2(0,T;H)$ it holds

$$\mathbf{E}\left[\left(\int_0^t \Phi_1(\sigma)\,\mathrm{d}W(\sigma), \int_0^s \Phi_2(\sigma)\,\mathrm{d}W(\sigma)\right)\right]$$

$$= \int_0^{t \wedge s} \mathbf{E}\left[(\Phi_1(\sigma), \Phi_2(\sigma))_{L_2^0}\right]\mathrm{d}\sigma$$

$$= \int_0^{t \wedge s} \mathbf{E}\left[\mathrm{Tr}\left(\Phi_1(\sigma)Q^{\frac{1}{2}}(\Phi_2(\sigma)Q^{\frac{1}{2}})^*\right)\right]\mathrm{d}\sigma.$$

Proposition 2.11 ([61, Lem. 2.4.1]). *For $\Phi \in \mathcal{N}_W^2(0, T; H)$ and $A \in L(H, H_1)$, where H_1 denotes a further Hilbert space $(H_1, (\cdot, \cdot)_{H_1}, \|\cdot\|_{H_1})$, it holds that $A \circ \Phi \in \mathcal{N}_W^2(0, T; H_1)$ and*

$$A \int_0^T \Phi(\sigma)\, dW(\sigma) = \int_0^T A\Phi(\sigma)\, dW(\sigma) \quad \textbf{P}\text{-}a.s.$$

The next lemma contains a *Burkholder–Davis–Gundy*-type inequality and is a special case of [18, Lem. 7.2]. It will be needed to estimate higher moments of stochastic integrals.

Proposition 2.12 ([18, Lem. 7.2]). *For any $p \geq 2$, $0 \leq \tau_1 < \tau_2 \leq T$, and for any predictable stochastic process $\Phi \colon [0, T] \times \Omega \to L_2^0$, which satisfies*

$$\mathbf{E}\left[\left(\int_{\tau_1}^{\tau_2} \|\Phi(\sigma)\|_{L_2^0}^2\, d\sigma\right)^{\frac{p}{2}}\right] < \infty,$$

we have

$$\mathbf{E}\left[\left\|\int_{\tau_1}^{\tau_2} \Phi(\sigma)\, dW(\sigma)\right\|^p\right] \leq C(p)\mathbf{E}\left[\left(\int_{\tau_1}^{\tau_2} \|\Phi(\sigma)\|_{L_2^0}^2\, d\sigma\right)^{\frac{p}{2}}\right].$$

Here the constant can be chosen to be

$$C(p) = \left(\frac{p}{2}(p-1)\right)^{\frac{p}{2}} \left(\frac{p}{p-1}\right)^{p(\frac{p}{2}-1)}.$$

We close this section with a remark on stochastic Itô-integrals with respect to cylindrical Wiener processes. Following [61, Chap. 2.5.2] let W be a Wiener process whose covariance operator $Q \in L(U)$ is not necessarily of finite trace. As in Proposition 2.7 we consider a further Hilbert space $(U_1, (\cdot, \cdot)_{U_1}, \|\cdot\|_{U_1})$ and a Hilbert–Schmidt embedding $J \colon U_0 \to U_1$ such that W is a standard Q_1-Wiener process on U_1 with $Q_1 = JJ^*$.

Consequently, by what is already established for standard Wiener processes, a predictable stochastic process Ψ is integrable with respect to the cylindrical Wiener process W if it takes values in $L_2(Q_1^{\frac{1}{2}}(U_1), H)$ and if

$$\mathbf{E}\left[\int_0^T \|\Psi(\sigma)\|_{L_2(Q_1^{\frac{1}{2}}(U_1), H)}^2\, d\sigma\right] < \infty.$$

Next, from Proposition 2.7 we get $Q_1^{\frac{1}{2}}(U_1) = J(U_0)$ and, by the polarization identity and (2.4), it holds for all $u_0, v_0 \in U_0$ that

$$\left(J u_0, J v_0 \right)_{Q_1^{\frac{1}{2}}(U_1)} = \frac{1}{4} \left(\left\| J(u_0 + v_0) \right\|^2_{Q_1^{\frac{1}{2}}(U_1)} + \left\| J(u_0 - v_0) \right\|^2_{Q_1^{\frac{1}{2}}(U_1)} \right)$$

$$= \frac{1}{4} \left(\left\| u_0 + v_0 \right\|^2_{U_0} + \left\| u_0 - v_0 \right\|^2_{U_0} \right)$$

$$= \left(u_0, v_0 \right)_{U_0}.$$

Hence, by following the lines of [61, Chap. 2.5.2] further, we obtain that $(J e_i)_{i \in \mathbb{N}}$ is an orthonormal basis of $Q_1^{\frac{1}{2}}(U_1)$ if $(e_i)_{i \in \mathbb{N}}$ is an orthonormal basis of U_0. This yields that $\Phi \in L_2^0$ if and only if $\Phi \circ J^{-1} \in L_2(Q_1^{\frac{1}{2}}(U_1), H)$. In fact, it holds

$$\| \Phi \|^2_{L_2^0} = \sum_{i=1}^{\infty} \| \Phi e_i \|^2 = \sum_{i=1}^{\infty} \left\| \Phi \circ J^{-1}(J e_i) \right\|^2 = \left\| \Phi \circ J^{-1} \right\|^2_{L_2(Q_1^{\frac{1}{2}}(U_1), H)}.$$

Hence, by defining

$$\int_0^T \Phi(\sigma) \, dW(\sigma) := \int_0^T \Phi(\sigma) \circ J^{-1} \, dW(\sigma) \tag{2.13}$$

the stochastic Itô-integral with respect to a cylindrical Wiener process is also well-defined for all $\Phi \in \mathcal{N}_W^2(0, T; H)$.

Finally, as in [61, Rem. 2.5.3], by considering (2.13) for elementary integrands (2.6) it can be seen that the definition (2.13) is independent of the particular choice of J and U_1.

2.3   Mild Solutions to Semilinear Stochastic Evolution Equations

In this section we introduce the semilinear stochastic evolution equation, which builds the starting point of this book. The properties of its solution are studied in the following sections. Here we focus on the formulation of the equation and we present our main assumptions. We also include several examples, some of them are studied more detailed in Chap. 6.

As in [18, Chap. 7.1] let $T > 0$ be a real number and let $(H, (\cdot, \cdot), \| \cdot \|)$ and $(U, (\cdot, \cdot)_U, \| \cdot \|_U)$ be separable Hilbert spaces. By $(\Omega, \mathscr{F}, \mathbf{P})$ let us denote a complete probability space and by $W : [0, T] \times \Omega \to U_1$ a cylindrical Q-Wiener process on a Hilbert space $U_1 \supset U$ adapted to a normal filtration $(\mathscr{F}_t)_{t \in [0, T]}$.

In this situation we are interested in a predictable stochastic process $X : [0, T] \times \Omega \to H$ which solves the following semilinear stochastic evolution equation (SEEq) driven by the Wiener process W

$$dX(t) + \big[AX(t) + f(t, X(t))\big]\,dt = g(t, X(t))\,dW(t), \qquad \text{for } 0 \le t \le T,$$
$$X(0) = X_0. \tag{2.14}$$

In the following we will discuss the meaning and the properties of all appearing objects in (2.14) and the given assumptions are sufficient for all results in this chapter. For some results the assumptions are easily relaxed to more general situations, but all details of possible generalizations are collected in Sect. 2.7.

Before we continue let us note that our framework and the assumptions are closely related to [18, Chap. 7.1]. The first assumption is concerned with the linear operator A and is more restrictive as its counter-part in [18, Chap. 7.1].

Assumption 2.13. The linear operator $A \colon \mathrm{dom}(A) \subset H \to H$ is densely defined, self-adjoint and positive definite with compact inverse.

As it is pointed out in Appendix B.2, the operator $-A$ is the generator of an analytic semigroup $(E(t))_{t \in [0,\infty)}$ on H and we obtain the existence of an increasing sequence of real numbers $(\lambda_i)_{i \in \mathbb{N}}$ with $\lim_{i \in \mathbb{N}} \lambda_i = \infty$ and an orthonormal basis of eigenvectors $(e_i)_{i \in \mathbb{N}}$ in H such that $Ae_i = \lambda_i e_i$ for all $i \in \mathbb{N}$.

We also recall the notion of fractional powers $A^{\frac{r}{2}}$ of the linear operator A from (B.1) and their domains which are denoted by $\dot{H}^r := \mathrm{dom}(A^{\frac{r}{2}})$ for all $r \in \mathbb{R}$. Note that $\dot{H}^r$ endowed with the inner product $(\cdot, \cdot)_r := (A^{\frac{r}{2}} \cdot, A^{\frac{r}{2}} \cdot)$ and the induced norm $\| \cdot \|_r$ is again a separable Hilbert space.

Assumption 2.14. The mapping $f \colon [0, T] \times \Omega \times H \to \dot{H}^{-1}$, $(t, \omega, h) \mapsto f(t, \omega, h)$ is $\mathscr{P}_T \times \mathscr{B}(H)/\mathscr{B}(\dot{H}^{-1})$-measurable.

Further, there exists a constant $C > 0$ such that $\|f(0, \omega, 0)\|_{-1} \le C$ for all $\omega \in \Omega$, and

$$\|f(t, \omega, h_1) - f(t, \omega, h_2)\|_{-1} \le C \|h_1 - h_2\| \tag{2.15}$$

for all $h_1, h_2 \in H$, $\omega \in \Omega$, $t \in [0, T]$. Also, there exists a constant $C > 0$ with

$$\|f(t_1, \omega, h) - f(t_2, \omega, h)\|_{-1} \le C(1 + \|h\|)(t_2 - t_1)^{\frac{1}{2}} \tag{2.16}$$

for all $h \in H$, $0 \le t_1 < t_2 \le T$, $\omega \in \Omega$.

The next two assumptions are concerned with the noise part of (2.14).

Assumption 2.15. The covariance operator $Q \in L(U)$ is self-adjoint and positive semidefinite but not necessarily of finite trace.

Proposition 2.6 yields the existence of a unique self-adjoint and positive definite operator $Q^{\frac{1}{2}} \in L(U)$ such that $Q^{\frac{1}{2}} \circ Q^{\frac{1}{2}} = Q$ and we recall the definition of the Hilbert space $U_0 = Q^{\frac{1}{2}}(U)$ from (2.3).

For the next assumption we also recall the notation $L_2^0 := L_2(U_0, H)$ for the set of all Hilbert–Schmidt operators which map from U_0 into H (see Definition 2.8).

Assumption 2.16. The mapping $g \colon [0, T] \times \Omega \times H \to L_2^0$, $(t, \omega, h) \mapsto g(t, \omega, h)$ is $\mathscr{P}_T \times \mathscr{B}(H)/\mathscr{B}(L_2^0)$-measurable.

Further, there exists a constant $C > 0$ such that $\|g(0, \omega, 0)\|_{L_2^0} \le C$ for all $\omega \in \Omega$, and

$$\|g(t, \omega, h_1) - g(t, \omega, h_2)\|_{L_2^0} \le C \|h_1 - h_2\| \tag{2.17}$$

for all $h_1, h_2 \in H$, $\omega \in \Omega$, $t \in [0, T]$. Also, there exists a constant $C > 0$ with

$$\|g(t_1, \omega, h) - g(t_2, \omega, h)\|_{L_2^0} \le C(1 + \|h\|)|t_1 - t_2|^{\frac{1}{2}} \tag{2.18}$$

for all $h \in H$, $t_1, t_2 \in [0, T]$, $\omega \in \Omega$.

As it is customary in stochastic analysis we often skip the explicit dependency of f and g and any other random variable with respect to $\omega \in \Omega$.

Finally, we impose the following measurability and regularity condition on X_0. For this let $r \in [0, 1]$ be given.

Assumption 2.17. The initial value $X_0 \colon \Omega \;\rightarrow\; \dot{H}^{1+r}$ is an $\mathscr{F}_0 / \mathscr{B}(\dot{H}^{1+r})$-measurable random variable and for some $p \ge 2$ it holds that

$$\|X_0\|_{L^p(\Omega; \dot{H}^{1+r})} < \infty.$$

Here, for an arbitrary Banach space B, we denote the norm in $L^p(\Omega, \mathscr{F}, \mathbf{P}; B)$ by $\| \cdot \|_{L^p(\Omega; B)}$, that is

$$\|Y\|_{L^p(\Omega; B)} = \left(\mathbf{E}\big[\|Y\|_B^p\big]\right)^{\frac{1}{p}}$$

for $Y \in L^p(\Omega, \mathscr{F}, \mathbf{P}; B)$.

Having this established we are in a position to define the notion of a mild solution to (2.14).

Definition 2.18. Let $p \ge 2$. A predictable stochastic process $X \colon [0, T] \times \Omega \rightarrow H$ is called a *p-fold integrable mild solution* to (2.14) if

$$\sup_{t \in [0, T]} \|X(t)\|_{L^p(\Omega; H)} < \infty \tag{2.19}$$

and, for all $t \in [0, T]$, it holds that

$$X(t) = E(t)X_0 - \int_0^t E(t - \sigma) f(\sigma, X(\sigma)) \, d\sigma$$
$$+ \int_0^t E(t - \sigma) g(\sigma, X(\sigma)) \, dW(\sigma) \quad \textbf{P}\text{-a.s.} \tag{2.20}$$

Equation (2.20) is often called *variation of constants formula* or *Duhamel's principle* for semilinear stochastic evolution equations.

We remark that Definition 2.18 is slightly more restrictive than the notion of mild solutions in [18, Chap. 7.1], where instead of (2.19) only the square integrability of the trajectories is required. But as it is shown in Sect. 2.4 our assumptions are also sufficient to ensure the existence of all integrals in (2.20) as p-fold integrable stochastic processes.

Further we note that in the Bochner integral part of (2.20) we in fact consider the extension of the semigroup $(E(t))_{t \in [0,T]}$ to $\dot{H}^{-1}$ in the sense of (B.5).

While the forgoing assumptions are sufficient to ensure the existence of a unique mild solution to (2.14) we work under the following two assumptions in order to derive the optimal regularity results in Sects. 2.5 and 2.6.

Let a parameter $r \in [0, 1]$ be given. As in [41] this parameter will determine the spatial regularity of the mild solution. Note that the linear growth conditions, especially for the mapping g, are easier to verify than the corresponding Lipschitz conditions.

Assumption 2.19. The mapping $f: [0, T] \times \Omega \times H \to \dot{H}^{-1}$ satisfies $f(t, \omega, h) \in \dot{H}^{-1+r}$ and

$$\left\| f(t, \omega, h) \right\|_{-1+r} \leq C \left(1 + \|h\|_r \right) \tag{2.21}$$

for all $h \in \dot{H}^r$, $t \in [0, T]$ and $\omega \in \Omega$.

In order to formulate the analogous assumption on g we introduce the notation $L_{2,r}^0$ which denotes the set of all Hilbert–Schmidt operators $\Phi: U_0 \to \dot{H}^r$ together with the norm $\|\Phi\|_{L_{2,r}^0} := \|A^{\frac{r}{2}} \Phi\|_{L_2^0}$.

Assumption 2.20. The mapping $g: [0, T] \times \Omega \times H \to L_2^0$ satisfies $g(t, \omega, h) \in L_{2,r}^0$ and

$$\left\| g(t, \omega, h) \right\|_{L_{2,r}^0} \leq C \left(1 + \|h\|_r \right) \tag{2.22}$$

for all $h \in \dot{H}^r$, $t \in [0, T]$ and $\omega \in \Omega$.

For $r = 0$ Assumptions 2.19 and 2.20 follow from Assumptions 2.14 and 2.16, respectively.

Example 2.21 (Stochastic heat equation). Our first example is the *stochastic heat equation* with additive noise on the unit interval. Let $H = L^2([0, 1], \mathscr{B}([0, 1]), d\xi; \mathbb{R})$ and consider a Q-Wiener process W on H with $\mathrm{Tr}(Q) < \infty$. The problem is to find a measurable mapping $X: [0, T] \times \Omega \times [0, 1] \to \mathbb{R}$ such that

$$dX(t, \xi) = \frac{d^2}{d\xi^2} X(t, \xi)\, dt + dW(t, \xi) \qquad \text{for all } t \in (0, T], \xi \in [0, 1],$$

$$X(t, 0) = X(t, 1) = 0, \qquad\qquad\qquad \text{for all } t \in (0, T], \tag{2.23}$$

$$X(0, \xi) = X_0(\xi), \qquad\qquad\qquad\quad \text{for all } \xi \in (0, 1),$$

where the initial condition X_0 is a random variable $X_0 \colon \Omega \times [0, 1] \to \mathbb{R}$ such that, for almost all $\omega \in \Omega$, $X_0(\omega, \cdot)$ is a sufficiently smooth function which also satisfies the boundary conditions.

We first translate this *stochastic partial differential equation* (SPDE) into an abstract stochastic evolution equation on the Hilbert space H. For this we understand the second order derivative with respect to ξ in (2.23) in the sense of *weak derivatives* of Sobolev spaces [25, Chap. 5.2] which give rise to the definition of an unbounded operator $A = -\frac{d^2}{d\xi^2}$ on $\operatorname{dom}(A) = H_0^1(0, 1) \cap H^2(0, 1)$. Here, by $H_0^1(0, 1)$ and $H^2(0, 1)$ we denote *Sobolev spaces* of square integrable functions on $[0, 1]$ with existing first and second order weak derivatives respectively, where the elements of $H_0^1(0, 1)$ also satisfy the homogeneous boundary conditions. For more details on Sobolev spaces we refer to [25, Chap. 5.2] and the references therein. From [25, Chap. 7.4, Th. 5] it follows that $-A$ is the generator of an analytic semigroup $(E(t))_{t \in [0,T]}$ on H.

Thus, we can rewrite (2.23) as an SEEq of the form

$$dX(t) + AX(t)\, dt = dW(t), \quad \text{for all } t \in [0, T],$$

$$X(0) = X_0,$$

where we are now looking for a stochastic process $X \colon [0, T] \times \Omega \to H$ as a solution to the SEEq.

The mild solution is then given by

$$X(t) = E(t)X_0 + \int_0^t E(t - \sigma)\, dW(\sigma), \quad t \in [0, T],$$

where the stochastic integral is often called *stochastic convolution*.

Example 2.22 (More general elliptic generators). Let $H := L^2(\mathscr{D}, \mathscr{B}(\mathscr{D}), dx\, ; \mathbb{R})$, where $\mathscr{D} \subset \mathbb{R}^d$, $d \in \mathbb{N}$, is a bounded domain with smooth boundary $\partial \mathscr{D}$ or a convex domain with polygonal boundary. Let the operator A be given by

$$Au := -\nabla \cdot (a(x)\nabla u) + c(x)u$$

with Dirichlet boundary conditions, where $a, c \colon \mathscr{D} \to \mathbb{R}$ are sufficiently smooth with $c(x) \geq 0$ and $a(x) \geq a_0 > 0$ for all $x \in \mathscr{D}$. Under this condition A satisfies Assumption 2.13, which can be shown by the same methods used in [53, Chap. 6.1].

Further, in this case it is well-known (for example, [53, Theorem 6.4] and [66, Chap. 3]) that $\dot{H}^1 = H_0^1(\mathscr{D})$ and $\dot{H}^2 = H^2(\mathscr{D}) \cap H_0^1(\mathscr{D})$, where $H^k(\mathscr{D})$, $k \geq 0$, denotes the Sobolev space of order k and $H_0^1(\mathscr{D})$ consists of all functions in $H^1(\mathscr{D})$ which are zero on the boundary. Furthermore, the norms in $H^k(\mathscr{D})$ and $\dot{H}^k$ are equivalent in $\dot{H}^k$ for $k = 1, 2$ (see [66, Lemma 3.1]).

We can also consider more general elliptic operators A of the form

$$Au := -\nabla \cdot (a(x)\nabla u) + b(x) \cdot \nabla u + c(x)u,$$

again with Dirichlet boundary conditions and $a, c: \mathscr{D} \to \mathbb{R}$, $b: \mathscr{D} \to \mathbb{R}^d$ sufficiently smooth with $a(x) \geq a_0 > 0$. In this case, A may not be self-adjoint and positive definite.

However, by splitting A into two parts, that is

$$A_1 u := -\nabla \cdot (a(x)\nabla u) \quad \text{and } f(u) := b(x) \cdot \nabla u + c(x)u,$$

we obtain a self-adjoint and positive definite part A_1 and a linear mapping $f: H \to \dot{H}^{-1}$. In fact, f satisfies

$$\int_{\mathscr{D}} f(u)v \, dx = \int_{\mathscr{D}} \sum_{j=1}^{d} b_j(x)\frac{\partial}{\partial x_j}u(x)v(x) + c(x)u(x)v(x) \, dx$$

$$= \int_{\mathscr{D}} -u(x) \sum_{j=1}^{d} \frac{\partial}{\partial x_j}\big(b_j(x)v(x)\big) + c(x)u(x)v(x) \, dx$$

$$\leq C \|u\|_{L^2(\mathscr{D})} \|v\|_{H_0^1}$$

for all $v \in H_0^1(\mathscr{D})$. Here the constant may be chosen to be

$$C = \sup_{x \in \mathscr{D}} \Big(\sum_{j=1}^{d} |b_j(x)| + \sum_{j=1}^{d} \Big|\frac{\partial}{\partial x_j}b_j(x)\Big| + |c(x)| \Big).$$

This way A_1 and f satisfy Assumptions 2.13 and 2.14.

Example 2.23 (Nemytskii operator). In this example we follow [41, Sect. 4] by describing a class of Hilbert–Schmidt operators which satisfy Assumptions 2.16 and 2.20. As in Example 2.21, let us consider $H = L^2([0, 1], \mathscr{B}([0, 1]), \, d\xi; \mathbb{R})$ and a trace class covariance operator Q on $U = H$, whose eigenfunctions $(\varphi_j)_{j \in \mathbb{N}}$ are uniformly bounded, that is

$$\sup_{j \in \mathbb{N}} \sup_{x \in [0,1]} |\varphi_j(x)| \leq C < \infty.$$

Further, let $\gamma: \mathbb{R} \to \mathbb{R}$ be a mapping satisfying

$$|\gamma(x_1) - \gamma(x_2)| \leq C|x_1 - x_2| \quad \text{for all } x_1, x_2 \in \mathbb{R}.$$

We then define $g\colon H \to L_2^0$ by

$$(g(u)v)[x] := \gamma(u(x))v(x), \quad \text{for all } x \in [0, 1],$$

and for all $v \in U_0 = Q^{\frac{1}{2}}(H)$. Indeed as in [41, Sect. 4], if we denote the eigenvalues of Q by μ_j, $j = 1, 2, \ldots$, we have

$$\|g(u)\|_{L_2^0}^2 = \sum_{j=1}^{\infty} \|g(u)\sqrt{\mu_j}\varphi_j\|^2 = \sum_{j=1}^{\infty} \mu_j \int_0^1 |\gamma(u(x))\varphi_j(x)|^2 \, \mathrm{d}x$$

$$\leq C \sum_{j=1}^{\infty} \mu_j \int_0^1 |\gamma(u(x))|^2 \, \mathrm{d}x \leq C \operatorname{Tr}(Q)(1 + \|u\|^2).$$

Hence, g is well-defined and often called the *Nemytskii operator* or *superposition operator* induced by γ. A similar calculation shows that g is globally Lipschitz.

Further, under the additional assumption that

$$\sum_{j=1}^{\infty} \mu_j \sup_{x \in [0,1]} |\varphi_j'(x)|^2 < \infty$$

a more involved analysis of g in [41, Sect. 4] shows that

$$\|g(u)\|_{L_{2,r}^0} \leq C(1 + \|u\|_r), \quad \text{for all } u \in \dot{H}^r$$

for all $r \in (0, \frac{1}{2})$.

Example 2.24 (SODE). We consider the following simple *stochastic ordinary differential equation* (SODE): Let $H = U = \mathbb{R}$ and W is a real-valued standard Wiener process. Then the solution to

$$\mathrm{d}X(t) = \mathrm{d}W(t), \quad X(0) = 0, \tag{2.24}$$

is simply given by

$$X(t) = W(t), \quad t \in [0, T].$$

In order to write (2.24) as a semilinear stochastic evolution equation, let us introduce an artificial drift term by setting $A = \lambda$, $f(x) = -\lambda x$ for $\lambda > 0$. Then we consider

$$\mathrm{d}X(t) + [AX(t) + f(X(t))] \, \mathrm{d}t = \mathrm{d}W(t), \quad X(0) = 0,$$

with mild solution

$$X(t) = \lambda \int_0^t e^{-\lambda(t-\sigma)} X(\sigma) \, d\sigma + \int_0^t e^{-\lambda(t-\sigma)} \, dW(\sigma).$$

Since $X(t) = W(t)$ it holds

$$W(t) = \lambda \int_0^t e^{-\lambda(t-\sigma)} W(\sigma) \, d\sigma + \int_0^t e^{-\lambda(t-\sigma)} \, dW(\sigma) \quad \text{for all } t \in [0, T].$$
$$(2.25)$$

In Sect. 4.3 we will extend the validity of (2.25) to much more general noise and semigroups in the abstract Hilbert space setting. This example is also continued in Sect. 2.6.

2.4 Existence and Uniqueness of Mild Solutions

In this section we prove an existence and uniqueness result for the stochastic evolution equation (2.14). The given proof is a generalized and slightly simplified version of the proof of [41, Th. 1] and it also contains some regularity estimates, which turn out to be useful in order to obtain the optimal regularity results in Sects. 2.5 and 2.6.

A further related result for stochastic evolution equations of the form (2.14) is found in [18, Th. 7.4].

Theorem 2.25. *Let Assumptions 2.13 to 2.17 be satisfied for some $p \geq 2$ and with $r = 0$. Then there exists a unique p-fold integrable mild solution $X : [0, T] \times \Omega \to H$ to (2.14) such that for every $t \in [0, T]$ and every $s \in [0, 1)$ it holds that $\mathbf{P}(X(t) \in \dot{H}^s) = 1$ with*

$$\sup_{t \in [0,T]} \|X(t)\|_{L^p(\Omega; \dot{H}^s)} < \infty. \qquad (2.26)$$

In addition, for every $\delta \in (0, \frac{1}{2})$ there exists a constant $C > 0$ with

$$\|X(t_1) - X(t_2)\|_{L^p(\Omega;H)} \leq C |t_1 - t_2|^\delta \qquad (2.27)$$

for all $t_1, t_2 \in [0, T]$.

Here uniqueness is understood in the sense that if $Y : [0, T] \times \Omega \to H$ is a further predictable stochastic process which satisfies (2.20) than for every $t \in [0, T]$ it holds that

$$\mathbf{P}(X(t) = Y(t)) = 1.$$

In this case, X and Y are *stochastically equivalent* or Y is called a *modification* or *version* of X.

As in the proof of [18, Th. 7.4] and [41, Th. 1] we consider the space $\mathcal{H}_p$ of all predictable stochastic processes $Y : [0, T] \times \Omega \to H$ which satisfy

$$\|Y\|_{\mathcal{H}_p} := \left(\sup_{t \in [0,T]} \mathbf{E}\big[\|Y(t)\|^p \big] \right)^{\frac{1}{p}} < \infty. \tag{2.28}$$

Note that after identifying stochastic processes which are stochastically equivalent $(\mathcal{H}_p, \|\cdot\|_{\mathcal{H}_p})$ becomes a Banach space.

The following lemma gives a temporal Hölder continuity and spatial Lipschitz continuity result on the integrands of the convolution integrals.

Lemma 2.26. *Let Assumptions 2.14, and 2.16 hold. Then, there exists a constant $C > 0$ such that*

$$\big\| f(\tau_1, Y(\tau_1)) - f(\tau_2, Z(\tau_2)) \big\|_{L^p(\Omega; \dot{H}^{-1})} + \big\| g(\tau_1, Y(\tau_1)) - g(\tau_2, Z(\tau_2)) \big\|_{L^p(\Omega; L_2^0)}$$

$$\leq C \big(1 + \|Y(\tau_1)\|_{L^p(\Omega;H)} \big) |\tau_1 - \tau_2|^{\frac{1}{2}} + C \big\| Y(\tau_1) - Z(\tau_2) \big\|_{L^p(\Omega;H)}$$

for all $\tau_1, \tau_2 \in [0, T]$ and all $Y, Z \in \mathcal{H}_p$.
In particular, we have

$$\big\| f(\tau, Y(\tau)) \big\|_{L^p(\Omega; \dot{H}^{-1})} + \big\| g(\tau, Y(\tau)) \big\|_{L^p(\Omega; L_2^0)} \leq C \big(1 + \|Y(\tau)\|_{L^p(\Omega;H)} \big) \tag{2.29}$$

for all $\tau \in [0, T]$, $Y \in \mathcal{H}_p$.

Proof. The proof of the lemma is straightforward. We have

$$\big\| g(\tau_1, Y(\tau_1)) - g(\tau_2, Z(\tau_2)) \big\|_{L^p(\Omega; L_2^0)} \leq \big\| g(\tau_1, Y(\tau_1)) - g(\tau_2, Y(\tau_1)) \big\|_{L^p(\Omega; L_2^0)}$$

$$+ \big\| g(\tau_2, Y(\tau_1)) - g(\tau_2, Z(\tau_2)) \big\|_{L^p(\Omega; L_2^0)}.$$

For the first summand it holds by (2.18) that

$$\big\| g(\tau_1, Y(\tau_1)) - g(\tau_2, Y(\tau_1)) \big\|_{L^p(\Omega; L_2^0)}$$

$$= \left(\int_\Omega \big\| g(\tau_1, \omega, Y(\tau_1, \omega)) - g(\tau_2, \omega, Y(\tau_1, \omega)) \big\|_{L_2^0}^p \, d\mathbf{P}(\omega) \right)^{\frac{1}{p}}$$

$$\leq C |\tau_1 - \tau_2|^{\frac{1}{2}} \left(\int_\Omega \big(1 + \|Y(\tau_1, \omega)\| \big)^p \, d\mathbf{P}(\omega) \right)^{\frac{1}{p}}$$

$$\leq C \big(1 + \|Y(\tau_1)\|_{L^p(\Omega;H)} \big) |\tau_1 - \tau_2|^{\frac{1}{2}}.$$

The second summand is estimated similarly by making use of (2.17). We obtain

$$
\left\| g(\tau_2, Y(\tau_1)) - g(\tau_2, Z(\tau_2)) \right\|_{L^p(\Omega; L_2^0)}
$$

$$
= \left(\int_\Omega \left\| g(\tau_2, \omega, Y(\tau_1, \omega)) - g(\tau_2, \omega, Z(\tau_2, \omega)) \right\|_{L_2^0}^p \, d\mathbf{P}(\omega) \right)^{\frac{1}{p}}
$$

$$
\leq C \left(\int_\Omega \left\| Y(\tau_1, \omega) - Z(\tau_2, \omega) \right\|^p \, d\mathbf{P}(\omega) \right)^{\frac{1}{p}}
$$

$$
= C \left\| Y(\tau_1) - Z(\tau_2) \right\|_{L^p(\Omega; H)}.
$$

By Assumption 2.16 we further have $\left\| g(0,0) \right\|_{L^p(\Omega; L_2^0)} < \infty$. Therefore,

$$
\left\| g(\tau, Y(\tau)) \right\|_{L^p(\Omega; L_2^0)} \leq \left\| g(\tau, Y(\tau)) - g(0,0) \right\|_{L^p(\Omega; L_2^0)} + \left\| g(0,0) \right\|_{L^p(\Omega; L_2^0)}
$$

$$
\leq C \left(1 + \left\| Y(\tau) \right\|_{L^p(\Omega; H)} \right),
$$

where the generic constant C also depends on T and $\left\| g(0,0) \right\|_{L^p(\Omega; L_2^0)}$ but is independent of Y.

The proof of f works in the exact same way. $\qquad\square$

Proof (of Theorem 2.25). The proof follows the lines of [18, Th. 7.4] and [41, Th. 1]. We introduce the mapping $\mathcal{K} : \mathcal{H}_p \to \mathcal{H}_p$ given by

$$
\mathcal{K}(Y)(t) := E(t)X_0 - \int_0^t E(t - \sigma) f(\sigma, Y(\sigma)) \, d\sigma
$$

$$
+ \int_0^t E(t - \sigma) g(\sigma, Y(\sigma)) \, dW(\sigma) \tag{2.30}
$$

$$
=: \mathcal{K}_0(t) - \mathcal{K}_1(Y)(t) + \mathcal{K}_2(Y)(t)
$$

for $t \in [0, T]$ and $Y \in \mathcal{H}_p$. First, we show that $\mathcal{K}$ is well-defined.

From Assumption 2.17 it follows that

$$
(t, \omega) \mapsto \mathcal{K}_0(t, \omega) = E(t) X_0(\omega)
$$

is an adapted, pathwise continuous and, therefore, predictable stochastic process with

$$
\sup_{t \in [0,T]} \mathbf{E} \left[\left\| A^{\frac{s}{2}} E(t) X_0 \right\|^p \right] \leq \mathbf{E} \left[\left\| A^{\frac{s}{2}} X_0 \right\|^p \right] \leq \mathbf{E} \left[\left\| X_0 \right\|_{\dot{H}^1}^p \right] < \infty, \tag{2.31}
$$

where we also applied Lemma B.9(i) and the fact that $\| E(t) \| \leq 1$ for all $t \in [0, T]$. Hence, $\mathcal{K}_0 \in \mathcal{H}_p$ and $\mathcal{K}_0(t)$ takes almost surely values in $\dot{H}^s$ for every $s \in [0, 1]$ and every $t \in [0, T]$. In addition, for every $\delta \in (0, \frac{1}{2}]$ Lemma B.9(i) and (ii) yield

$$\|\mathcal{K}_0(t_1) - \mathcal{K}_0(t_2)\|_{L^p(\Omega;H)} = \|(E(t_2 - t_1) - \mathrm{Id}_H)E(t_1)X_0\|_{L^p(\Omega;H)}$$

$$\leq \|A^{-\delta}(E(t_2 - t_1) - \mathrm{Id}_H)\|\,\|E(t_1)A^{\delta}X_0\|_{L^p(\Omega;H)}$$

$$\leq C(t_2 - t_1)^{\delta}\|X_0\|_{L^p(\Omega;\dot{H}^{-1})} \tag{2.32}$$

for all $t_1, t_2 \in [0, T]$, $t_1 < t_2$.

Next, we prove similar estimates for the Bochner integral part in (2.30). From Lemma B.9(i) and (2.29) we obtain

$$\|\mathcal{K}_1(Y)(t)\|_{L^p(\Omega;\dot{H}^s)} = \left\|A^{\frac{s}{2}}\int_0^t E(t - \sigma)f(\sigma, Y(\sigma))\,d\sigma\right\|_{L^p(\Omega;H)}$$

$$= \int_0^t \left\|A^{\frac{s+1}{2}}E(t - \sigma)A^{-\frac{1}{2}}f(\sigma, Y(\sigma))\right\|_{L^p(\Omega;H)}\,d\sigma$$

$$\leq \int_0^t (t - \sigma)^{-\frac{s+1}{2}}\left\|A^{-\frac{1}{2}}f(\sigma, Y(\sigma))\right\|_{L^p(\Omega;H)}\,d\sigma \tag{2.33}$$

$$\leq C\int_0^t (t - \sigma)^{-\frac{s+1}{2}}\,d\sigma\left(1 + \sup_{\sigma\in[0,T]}\|Y(\sigma)\|_{L^p(\Omega;H)}\right)$$

$$\leq C\frac{2}{1 - s}T^{\frac{1-s}{2}}\left(1 + \|Y\|_{\mathcal{H}_p}\right).$$

Hence, for every $Y \in \mathcal{H}_p$ it holds that $\mathcal{K}_1(Y)$ is an adapted stochastic process such that

$$\mathbf{P}\left(\mathcal{K}_1(Y)(t) \in \dot{H}^s\right) = 1$$

for every $s \in [0, 1)$ and $t \in [0, T]$. Hence, we have

$$\sup_{t\in[0,T]}\left\|\mathcal{K}_1(Y)(t)\right\|_{L^p(\Omega;\dot{H}^s)} \leq C\left(1 + \|Y\|_{\mathcal{H}_p}\right) < \infty \tag{2.34}$$

and, for $0 \leq t_1 < t_2 \leq T$, it follows

$$\left\|\mathcal{K}_1(Y)(t_1) - \mathcal{K}_1(Y)(t_2)\right\|_{L^p(\Omega;H)}$$

$$= \left\|\int_0^{t_1} E(t_1 - \sigma)f(\sigma, Y(\sigma))\,d\sigma - \int_0^{t_2} E(t_2 - \sigma)f(\sigma, Y(\sigma))\,d\sigma\right\|_{L^p(\Omega;H)}$$

$$\leq \left\|\int_{t_1}^{t_2} E(t_2 - \sigma)f(\sigma, Y(\sigma))\,d\sigma\right\|_{L^p(\Omega;H)}$$

$$+ \left\|(E(t_2 - t_1) - \mathrm{Id}_H)\int_0^{t_1} E(t_1 - \sigma)f(\sigma, Y(\sigma))\,d\sigma\right\|_{L^p(\Omega;H)}.$$

The first summand is estimated in the same way as (2.33) with $s = 0$ which reads

$$\left\| \int_{t_1}^{t_2} E(t_2 - \sigma) f(\sigma, Y(\sigma)) \, d\sigma \right\|_{L^p(\Omega;H)}$$

$$\leq C \int_{t_1}^{t_2} (t_2 - \sigma)^{-\frac{1}{2}} \, d\sigma \left(1 + \sup_{\sigma \in [0,T]} \| Y(\sigma) \|_{L^p(\Omega;H)} \right)$$

$$\leq C(t_2 - t_1)^{\frac{1}{2}} \left(1 + \| Y \|_{\mathscr{H}_p} \right).$$

By an application of Lemma B.9(*ii*) we obtain for the second summand and for every $\delta \in (0, \frac{1}{2})$

$$\left\| (E(t_2 - t_1) - \mathrm{Id}_H) \int_0^{t_1} E(t_1 - \sigma) f(\sigma, Y(\sigma)) \, d\sigma \right\|_{L^p(\Omega;H)}$$

$$\leq C(t_2 - t_1)^{\delta} \left\| A^{\delta} \int_0^{t_1} A^{\frac{1}{2}} E(t_1 - \sigma) A^{-\frac{1}{2}} f(\sigma, Y(\sigma)) \, d\sigma \right\|_{L^p(\Omega;H)}$$

$$\leq C(t_2 - t_1)^{\delta} \frac{2}{1 - 2\delta} T^{\frac{1}{2} - \delta} \left(1 + \| Y \|_{\mathscr{H}_p} \right).$$

Altogether, this completes the proof of

$$\| \mathscr{K}_1(Y)(t_1) - \mathscr{K}_1(Y)(t_2) \| \leq C(t_2 - t_1)^{\delta} \tag{2.35}$$

for all $0 \leq t_1 < t_2 \leq T$. From this and by Da Prato and Zabczyk [18, Prop. 3.6(ii)] we conclude that for every $Y \in \mathscr{H}_p$ the stochastic process $\mathscr{K}_1(Y)$ has a predictable version and, hence, $\mathscr{K}_1(Y) \in \mathscr{H}_p$.

In the following we show that similar estimates also hold for the stochastic integral $\mathscr{K}_2(Y)$ in (2.30). By Lemma B.9(*i*) and (2.29) we have for all $s \in [0, 1)$ and $0 \leq \tau_1 < \tau_2 \leq T$

$$\left(\mathbf{E} \left[\left(\int_{\tau_1}^{\tau_2} \| A^{\frac{s}{2}} E(\tau_2 - \sigma) g(\sigma, Y(\sigma)) \|_{L_2^0}^2 \, d\sigma \right)^{\frac{p}{2}} \right] \right)^{\frac{1}{p}}$$

$$\leq \left(\mathbf{E} \left[\left(\int_{\tau_1}^{\tau_2} (\tau_2 - \sigma)^{-s} \| g(\sigma, Y(\sigma)) \|_{L_2^0}^2 \, d\sigma \right)^{\frac{p}{2}} \right] \right)^{\frac{1}{p}} \tag{2.36}$$

$$\leq \left(\int_{\tau_1}^{\tau_2} (\tau_2 - \sigma)^{-s} \| g(\sigma, Y(\sigma)) \|_{L^p(\Omega;L_2^0)}^2 \, d\sigma \right)^{\frac{1}{2}}$$

$$\leq C \frac{1}{(1 - s)^{\frac{1}{2}}} (\tau_2 - \tau_1)^{\frac{1 - s}{2}} \left(1 + \| Y \|_{\mathscr{H}_p} \right).$$

Further, under the given assumptions on g and Y and for fixed $t \in [0, T]$ the stochastic process $[0, t] \ni \sigma \mapsto E(t - \sigma) g(\sigma, Y(\sigma))$ is predictable and, because of (2.36) with $s = 0$, stochastically integrable. Therefore, for every fixed $t \in [0, T]$ the stochastic integral $\mathscr{K}_2(Y)(t)$ is a well-defined H-valued random variable which is $\mathscr{F}_t$-measurable.

Moreover, by applying Proposition 2.12 we get for every $s \in [0, 1)$

$$\left\| \mathscr{K}_2(Y)(t) \right\|_{L^p(\Omega;\dot{H}^s)} \leq C \left(\mathbf{E} \left[\left(\int_0^t \left\| A^{\frac{s}{2}} E(t - \sigma) g(\sigma, Y(\sigma)) \right\|_{L_2^0}^2 d\sigma \right)^{\frac{p}{2}} \right] \right)^{\frac{1}{p}}$$

and, consequently, by (2.36)

$$\sup_{t \in [0,T]} \left\| \mathscr{K}_2(Y)(t) \right\|_{L^p(\Omega;\dot{H}^s)} \leq C \left(1 + \| Y \|_{\mathscr{H}_p} \right) < \infty. \tag{2.37}$$

Therefore, for every $t \in [0, T]$ the random variable $\mathscr{K}_2(Y)(t)$ takes almost surely values in $\dot{H}^s$ for all $s \in [0, 1)$.

Further, for all $0 \leq t_1 < t_2 \leq T$ and $\delta \in (0, \frac{1}{2})$ it follows by Proposition 2.12, Lemma B.9(ii) and (2.36) that

$$\left\| \mathscr{K}_2(Y)(t_1) - \mathscr{K}_2(Y)(t_2) \right\|_{L^p(\Omega;H)}$$

$$\leq \left\| \int_{t_1}^{t_2} E(t_2 - \sigma) g(\sigma, Y(\sigma)) \, dW(\sigma) \right\|_{L^p(\Omega;H)}$$

$$+ \left\| \left(E(t_2 - t_1) - \mathrm{Id}_H \right) \int_0^{t_1} E(t_1 - \sigma) g(\sigma, Y(\sigma)) \, dW(\sigma) \right\|_{L^p(\Omega;H)} \tag{2.38}$$

$$\leq C \left(\mathbf{E} \left[\left(\int_{t_1}^{t_2} \left\| E(t_2 - \sigma) g(\sigma, Y(\sigma)) \right\|_{L_2^0}^2 d\sigma \right)^{\frac{p}{2}} \right] \right)^{\frac{1}{p}}$$

$$+ C (t_2 - t_1)^{\delta} \left(\mathbf{E} \left[\left(\int_0^{t_1} \left\| A^{\delta} E(t_1 - \sigma) g(\sigma, Y(\sigma)) \right\|_{L_2^0}^2 d\sigma \right)^{\frac{p}{2}} \right] \right)^{\frac{1}{p}}$$

$$\leq C \left(1 + \frac{1}{(1 - 2\delta)^{\frac{1}{2}}} \right) T^{\frac{1-2\delta}{2}} \left(1 + \| Y \|_{\mathscr{H}_p} \right) (t_2 - t_1)^{\delta}.$$

Hence, by applying [18, Prop. 3.6(ii)] we again conclude that for every $Y \in \mathscr{H}_p$ the stochastic process $\mathscr{K}_2(Y)$ has a predictable version and, hence, $\mathscr{K}_2(Y) \in \mathscr{H}_p$. Combining all results for $\mathscr{K}_0$, $\mathscr{K}_1(Y)$ and $\mathscr{K}_2(Y)$ shows that the mapping $\mathscr{K}$ indeed maps into $\mathscr{H}_p$.

It remains to show that $\mathscr{K}$ is a contraction. For this, we further follow the lines of the proof of [41, Th. 1] and for $u \in \mathbb{R}$ we introduce the norm

$$\| Y \|_{\mathscr{H}_p, u} := \sup_{t \in [0,T]} e^{-ut} \| Y(t) \|_{L^p(\Omega;H)}$$

on $\mathscr{H}_p$ which is equivalent to $\| \cdot \|_{\mathscr{H}_p}$.

Then, for $Y, Z \in \mathscr{H}_p$ we get

$$\big\| \mathscr{K}(Y)(t) - \mathscr{K}(Z)(t) \big\|_{L^p(\Omega;H)}$$
$$\leq \big\| \mathscr{K}_1(Y)(t) - \mathscr{K}_1(Z)(t) \big\|_{L^p(\Omega;H)} + \big\| \mathscr{K}_2(Y)(t) - \mathscr{K}_2(Z)(t) \big\|_{L^p(\Omega;H)}.$$

Further, by Lemma 2.26 it holds

$$\big\| f(\sigma, Y(\sigma)) - f(\sigma, Z(\sigma)) \big\|_{L^p(\Omega;\dot{H}^{-1})} + \big\| g(\sigma, Y(\sigma)) - g(\sigma, Z(\sigma)) \big\|_{L^p(\Omega;L_2^0)}$$
$$\leq C \| Y(\sigma) - Z(\sigma) \|_{L^p(\Omega;H)}$$

$$(2.39)$$

for all $\sigma \in [0, T]$. Thus, for the first summand we have

$$\big\| \mathscr{K}_1(Y)(t) - \mathscr{K}_1(Z)(t) \big\|_{L^p(\Omega;H)}$$
$$\leq \left\| \int_0^t E(t-\sigma)\big(f(\sigma, Y(\sigma)) - f(\sigma, Z(\sigma))\big) \, d\sigma \right\|_{L^p(\Omega;H)}$$
$$\leq \int_0^t \big\| A^{\frac{1}{2}} E(t-\sigma) A^{-\frac{1}{2}} \big(f(\sigma, Y(\sigma)) - f(\sigma, Z(\sigma))\big) \big\|_{L^p(\Omega;H)} \, d\sigma$$
$$\leq C \int_0^t (t-\sigma)^{-\frac{1}{2}} \big\| f(\sigma, Y(\sigma)) - f(\sigma, Z(\sigma)) \big\|_{L^p(\Omega;\dot{H}^{-1})} \, d\sigma$$
$$\leq C \int_0^t (t-\sigma)^{-\frac{1}{2}} \big\| Y(\sigma) - Z(\sigma) \big\|_{L^p(\Omega;H)} \, d\sigma$$
$$\leq C \int_0^t (t-\sigma)^{-\frac{1}{2}} e^{u\sigma} \, d\sigma \, \| Y - Z \|_{\mathscr{H}_p,u},$$

where we also applied Lemma B.9(i). Analogously, we obtain for the second summand by Proposition 2.12

$$\big\| \mathscr{K}_2(Y)(t) - \mathscr{K}_2(Z)(t) \big\|_{L^p(\Omega;H)}$$
$$\leq C \left(\mathbf{E}\left[\left(\int_0^t \big\| E(t-\sigma)\big(g(\sigma, Y(\sigma)) - g(\sigma, Z(\sigma))\big) \big\|_{L_2^0}^2 \, d\sigma \right)^{\frac{p}{2}} \right] \right)^{\frac{1}{p}}$$
$$\leq C \left(\int_0^t \big\| g(\sigma, Y(\sigma)) - g(\sigma, Z(\sigma)) \big\|_{L^p(\Omega;L_2^0)}^2 \, d\sigma \right)^{\frac{1}{2}}$$
$$\leq C \left(\int_0^t e^{2u\sigma} \, d\sigma \right)^{\frac{1}{2}} \| Y - Z \|_{\mathscr{H}_p,u}.$$

Hence, combining both estimates yields

$$\left\| \mathscr{K}(Y)(t) - \mathscr{K}(Z)(t) \right\|_{L^p(\Omega;H)}$$

$$\leq C \left(\int_0^t (t-\sigma)^{-\frac{1}{2}} e^{u\sigma} \, d\sigma + \left(\frac{1}{2u} (e^{2ut} - 1) \right)^{\frac{1}{2}} \right) \|Y - Z\|_{\mathscr{H}_p, u}.$$

By Hölder's inequality with $1 = \frac{1}{q} + \frac{1}{q'}$ with $q = 3$ we also obtain

$$\int_0^t (t-\sigma)^{-\frac{1}{2}} e^{u\sigma} \, d\sigma \leq \left(\int_0^t (t-\sigma)^{-\frac{3}{4}} \, d\sigma \right)^{\frac{2}{3}} \left(\int_0^t e^{3u\sigma} \, d\sigma \right)^{\frac{1}{3}}$$

$$\leq 4^{\frac{2}{3}} T^{\frac{1}{6}} \left(\frac{1}{3u} (e^{3ut} - 1) \right)^{\frac{1}{3}}.$$

Hence, for all $u > 0$ we have

$$e^{-ut} \left\| \mathscr{K}(Y)(t) - \mathscr{K}(Z)(t) \right\|_{L^p(\Omega;H)}$$

$$\leq C \left(4^{\frac{2}{3}} T^{\frac{1}{6}} \left(\frac{1}{3u} (1 - e^{-3ut}) \right)^{\frac{1}{3}} + \left(\frac{1}{2u} (1 - e^{-2ut}) \right)^{\frac{1}{2}} \right) \|Y - Z\|_{\mathscr{H}_p, u}$$

$$\leq C \left(4^{\frac{2}{3}} T^{\frac{1}{6}} (3u)^{-\frac{1}{3}} + (2u)^{-\frac{1}{2}} \right) \|Y - Z\|_{\mathscr{H}_p, u}.$$

Therefore, for $u > 0$ sufficiently large $\mathscr{K} : \mathscr{H}_p \to \mathscr{H}_p$ is a contraction with respect to the norm $\| \cdot \|_{\mathscr{H}_p, u}$ and there exists a unique fixed point $X \in \mathscr{H}_p$, which is a unique mild solution to (2.14). The spatial regularity (2.26) follows from (2.31), (2.34) and (2.37) and the asserted Hölder continuity (2.27) of X follows from (2.32), (2.35) and (2.38). □

2.5 Optimal Spatial Regularity

The aim of this section is to prove the following spatial regularity result. At the end of this section we also discuss an example in which the results of the theorem are indeed optimal.

Theorem 2.27. *Let $r \in [0, 1)$, $p \in [2, \infty)$ be given such that Assumptions 2.13–2.17 and Assumptions 2.19 and 2.20 are fulfilled. Then the unique mild solution $X : [0, T] \times \Omega \to H$ to (2.14) satisfies*

$$\mathbf{P}\left(X(t) \in \dot{H}^{1+r} \right) = 1$$

for all $t \in [0, T]$. In addition, there exists a constant $C > 0$ such that

$$\sup_{t \in [0,T]} \|X(t)\|_{L^p(\Omega; \dot{H}^{1+r})} \leq \|X_0\|_{L^p(\Omega; \dot{H}^{1+r})} + C\left(1 + \sup_{t \in [0,T]} \|X(t)\|_{L^p(\Omega; \dot{H}^r)}\right).$$

(2.40)

Before we continue let us note that all appearing terms on the right hand side of (2.40) are finite by Assumption 2.17 and the results of Theorem 2.25.

The proof is divided in several mostly technical lemmas, but some of them have useful applications in others parts of this work.

The following two lemmas contain our main idea of proof and yield the key estimates. The applied technique is already known in the literature but only applied to Bochner integrals consisting of a convolution with an analytic semigroup, for example in [19, Prop. 3] and [65, p. 157]. To the best of our knowledge the application of this technique to stochastic integrals appeared for the first time in [50].

Lemma 2.28. *Let Assumptions 2.13, 2.14 and 2.19 hold with $r \in [0, 1)$ and $p \geq 2$. Given a predictable stochastic process $Y : [0, T] \times \Omega \to H$ with $\mathbf{P}(Y(t) \in \dot{H}^r) = 1$ for all $t \in [0, T]$ and*

$$\sup_{t \in [0,T]} \|Y(t)\|_{L^p(\Omega; \dot{H}^r)} < \infty.$$

Then for all $s \in [0, r + 1]$ there exists a constant $C = C(r, s, A, f)$ such that, for all $0 \leq \tau_1 < \tau_2 \leq T$,

$$\left\| A^{\frac{s}{2}} \int_{\tau_1}^{\tau_2} E(\tau_2 - \sigma) f(\tau_2, Y(\tau_2)) \, d\sigma \right\|_{L^p(\Omega; H)}$$

$$\leq C\left(1 + \sup_{\sigma \in [0,T]} \|Y(\sigma)\|_{L^p(\Omega; \dot{H}^r)}\right)(\tau_2 - \tau_1)^{\frac{1+r-s}{2}}.$$

(2.41)

If, in addition, for some $\delta > \frac{r}{2}$ there exists C_δ such that

$$\|Y(t_1) - Y(t_2)\|_{L^p(\Omega; H)} \leq C_\delta |t_2 - t_1|^\delta \text{ for all } t_1, t_2 \in [0, T],$$

then we also have, with $C = C(\delta, s, r, f, C_\delta)$, that

$$\left\| A^{\frac{s}{2}} \int_{\tau_1}^{\tau_2} E(\tau_2 - \sigma)\big(f(\tau_2, Y(\tau_2)) - f(\sigma, Y(\sigma))\big) \, d\sigma \right\|_{L^p(\Omega; H)}$$

$$\leq \frac{C}{1 + 2\delta - s}\left(1 + \sup_{\sigma \in [0,T]} \|Y(\sigma)\|_{L^p(\Omega; H)}\right)(\tau_2 - \tau_1)^{\frac{1+2\delta-s}{2}}.$$

(2.42)

In particular, with $C = C(T, \delta, r, s, f, C_\delta)$ *it holds that*

$$\left\| A^{\frac{s}{2}} \int_{\tau_1}^{\tau_2} E(\tau_2 - \sigma) f(\sigma, Y(\sigma)) \, d\sigma \right\|_{L^p(\Omega;H)}$$

$$\leq C \Big(1 + \sup_{\sigma \in [0,T]} \|Y(\sigma)\|_{L^p(\Omega;\dot{H}^r)} \Big) (\tau_2 - \tau_1)^{\frac{1+r-s}{2}}. \tag{2.43}$$

Proof. For $0 \leq \tau_1 < \tau_2 \leq T$ we have

$$\left\| A^{\frac{s}{2}} \int_{\tau_1}^{\tau_2} E(\tau_2 - \sigma) f(\sigma, Y(\sigma)) \, d\sigma \right\|_{L^p(\Omega;H)}$$

$$\leq \left\| A^{\frac{s}{2}} \int_{\tau_1}^{\tau_2} E(\tau_2 - \sigma) f(\tau_2, Y(\tau_2)) \, d\sigma \right\|_{L^p(\Omega;H)}$$

$$+ \left\| A^{\frac{s}{2}} \int_{\tau_1}^{\tau_2} E(\tau_2 - \sigma) \big(f(\tau_2, Y(\tau_2)) - f(\sigma, Y(\sigma)) \big) \, d\sigma \right\|_{L^p(\Omega;H)}.$$

Therefore, if we show (2.41) and (2.42) then (2.43) follows immediately by using $(\tau_2 - \tau_1)^{\delta - \frac{r}{2}} \leq T^{\delta - \frac{r}{2}}$.

For (2.41) first note that by Assumption 2.19 the random variable $A^{\frac{-1+r}{2}} F(Y(\tau_2))$ takes values in H almost surely. Hence, we can apply Lemma B.9*(iv)* and we obtain

$$\left\| A^{\frac{s}{2}} \int_{\tau_1}^{\tau_2} E(\tau_2 - \sigma) f(\tau_2, Y(\tau_2)) \, d\sigma \right\|_{L^p(\Omega;H)}$$

$$= \left\| A^{\frac{s+1-r}{2}} \int_{\tau_1}^{\tau_2} E(\tau_2 - \sigma) A^{\frac{-1+r}{2}} f(\tau_2, Y(\tau_2)) \, d\sigma \right\|_{L^p(\Omega;H)}$$

$$\leq C(r, s)(\tau_2 - \tau_1)^{\frac{1+r-s}{2}} \left\| A^{\frac{-1+r}{2}} f(\tau_2, Y(\tau_2)) \right\|_{L^p(\Omega;H)}$$

$$\leq C(r, s, f) \Big(1 + \sup_{\sigma \in [0,T]} \|Y(\sigma)\|_{L^p(\Omega;\dot{H}^r)} \Big) (\tau_2 - \tau_1)^{\frac{1+r-s}{2}}.$$

Finally, by Lemma B.2*(i)* and Lemma 2.26, we get (2.42):

$$\left\| A^{\frac{s}{2}} \int_{\tau_1}^{\tau_2} E(\tau_2 - \sigma) \big(f(\tau_2, Y(\tau_2)) - f(\sigma, Y(\sigma)) \big) \, d\sigma \right\|_{L^p(\Omega;H)}$$

$$\leq \int_{\tau_1}^{\tau_2} \left\| A^{\frac{s+1}{2}} E(\tau_2 - \sigma) A^{-\frac{1}{2}} \big(f(\tau_2, Y(\tau_2)) - f(\sigma, Y(\sigma)) \big) \right\|_{L^p(\Omega;H)} d\sigma$$

$$\leq C(r, s, f, C_\delta) \int_{\tau_1}^{\tau_2} (\tau_2 - \sigma)^{-\frac{s+1-2\delta}{2}} \big(1 + \|Y(\sigma)\|_{L^p(\Omega;H)} \big) \, d\sigma$$

$$= \frac{2C(r, s, f, C_\delta)}{1 + 2\delta - s} (\tau_2 - \tau_1)^{\frac{1+2\delta-s}{2}} \Big(1 + \sup_{\sigma \in [0,T]} \|Y(\sigma)\|_{L^p(\Omega;H)} \Big),$$

where we also used that $\frac{2\delta-s-1}{2} > -1$ since $s \le r + 1$ and $\delta > \frac{r}{2}$. This completes the proof. $\square$

Lemma 2.29. *Let Assumptions 2.13, 2.16 and 2.20 hold with $r \in [0, 1)$ and $p \ge 2$. Given a predictable stochastic process $Y : [0, T] \times \Omega \to H$ with $\mathbf{P}\big(Y(t) \in \dot{H}^r\big) = 1$ for all $t \in [0, T]$ and*

$$\sup_{t \in [0,T]} \|Y(t)\|_{L^p(\Omega; \dot{H}^r)} < \infty.$$

Then for all $s \in [0, r + 1]$ there exists a constant $C = C(r, s, A, g)$ such that, for all $0 \le \tau_1 < \tau_2 \le T$,

$$
\left(\mathbf{E}\left[\left(\int_{\tau_1}^{\tau_2} \left\| A^{\frac{s}{2}} E(\tau_2 - \sigma) g(\tau_2, Y(\tau_2)) \right\|_{L_2^0}^2 \, d\sigma \right)^{\frac{p}{2}} \right] \right)^{\frac{1}{p}}
$$
$$
\le C\left(1 + \sup_{t \in [0,T]} \|Y(t)\|_{L^p(\Omega; \dot{H}^r)}\right)(\tau_2 - \tau_1)^{\min\left(\frac{1}{2}, \frac{1+r-s}{2}\right)}.
\tag{2.44}
$$

If, in addition, for some $\frac{1}{2} \ge \delta > \frac{r}{2}$ there exists C_δ such that

$$\|Y(t_1) - Y(t_2)\|_{L^p(\Omega; H)} \le C_\delta |t_2 - t_1|^\delta \text{ for all } t_1, t_2 \in [0, T],$$

then we also have, with $C = C(s, A, g, C_\delta)$, that

$$
\left(\mathbf{E}\left[\left(\int_{\tau_1}^{\tau_2} \left\| A^{\frac{s}{2}} E(\tau_2 - \sigma) \big(g(\sigma, Y(\sigma)) - g(\tau_2, Y(\tau_2)) \big) \right\|_{L_2^0}^2 \, d\sigma \right)^{\frac{p}{2}} \right] \right)^{\frac{1}{p}}
$$
$$
\le \frac{C}{\sqrt{1 + 2\delta - s}}\left(1 + \sup_{t \in [0,T]} \|Y(t)\|_{L^p(\Omega; H)}\right)(\tau_2 - \tau_1)^{\frac{1+2\delta-s}{2}}.
\tag{2.45}
$$

In particular, with $C = C(T, \delta, p, r, s, A, g, C_\delta)$ it holds that

$$
\left\| \int_{\tau_1}^{\tau_2} A^{\frac{s}{2}} E(\tau_2 - \sigma) g(\sigma, Y(\sigma)) \, dW(\sigma) \right\|_{L^p(\Omega; H)}
$$
$$
\le C\left(1 + \sup_{t \in [0,T]} \|Y(t)\|_{L^p(\Omega; \dot{H}^r)}\right)(\tau_2 - \tau_1)^{\min\left(\frac{1}{2}, \frac{1+r-s}{2}\right)}.
\tag{2.46}
$$

Proof. As in the previous lemma the main idea is to use the Hölder continuity of Y to estimate the left-hand side in (2.46). First, let us fix $0 \le \tau_1 < \tau_2 \le T$. Then, the mapping $[\tau_1, \tau_2] \ni \sigma \mapsto A^{\frac{s}{2}} E(\tau_2 - \sigma) g(\sigma, Y(\sigma))$ is a predictable L_2^0-valued stochastic process. If we can show that

$$
\mathbf{E}\left[\left(\int_{\tau_1}^{\tau_2} \left\| A^{\frac{s}{2}} E(\tau_2 - \sigma) g(\sigma, Y(\sigma)) \right\|_{L_2^0}^2 \, d\sigma \right)^{\frac{p}{2}} \right] < \infty
\tag{2.47}
$$

the stochastic integral in (2.46) is well-defined and Proposition 2.12 is applicable. To this end,

$$\left(\mathbf{E}\left[\left(\int_{\tau_1}^{\tau_2}\left\|A^{\frac{s}{2}}E(\tau_2-\sigma)g(\sigma,Y(\sigma))\right\|_{L_2^0}^2\,d\sigma\right)^{\frac{p}{2}}\right]\right)^{\frac{1}{p}}$$

$$=\left\|\left(\int_{\tau_1}^{\tau_2}\left\|A^{\frac{s}{2}}E(\tau_2-\sigma)g(\sigma,Y(\sigma))\right\|_{L_2^0}^2\,d\sigma\right)^{\frac{1}{2}}\right\|_{L^p(\Omega;\mathbb{R})}$$

$$\leq\left\|\left(\int_{\tau_1}^{\tau_2}\left\|A^{\frac{s}{2}}E(\tau_2-\sigma)g(\tau_2,Y(\tau_2))\right\|_{L_2^0}^2\,d\sigma\right)^{\frac{1}{2}}\right\|_{L^p(\Omega;\mathbb{R})}$$

$$+\left\|\left(\int_{\tau_1}^{\tau_2}\left\|A^{\frac{s}{2}}E(\tau_2-\sigma)\big(g(\sigma,Y(\sigma))-g(\tau_2,Y(\tau_2))\big)\right\|_{L_2^0}^2\,d\sigma\right)^{\frac{1}{2}}\right\|_{L^p(\Omega;\mathbb{R})}$$

$$=:S_1+S_2.$$

In the second step we applied the triangle inequality. Now we deal with both summands separately.

In the term S_1 the time in $g(\tau_2, Y(\tau_2))$ is fixed. We also notice that $\eta := s - r - \max(0, s - r) \leq 0$ and, hence, $A^{\frac{\eta}{2}}$ is a bounded linear operator on H. Furthermore, since $s \in [0, r + 1]$ we have $\rho := \max(0, s - r) \in [0, 1]$ and Lemma B.9(iii) is applicable. By writing $s = \eta + \rho + r$, we get

$$\int_{\tau_1}^{\tau_2}\left\|A^{\frac{s}{2}}E(\tau_2-\sigma)g(\tau_2,Y(\tau_2))\right\|_{L_2^0}^2\,d\sigma$$

$$=\int_{\tau_1}^{\tau_2}\sum_{m=1}^{\infty}\left\|A^{\frac{s}{2}}E(\tau_2-\sigma)g(\tau_2,Y(\tau_2))\varphi_m\right\|^2\,d\sigma$$

$$\leq\sum_{m=1}^{\infty}\int_{\tau_1}^{\tau_2}\left\|A^{\frac{\eta}{2}}\right\|^2\left\|A^{\frac{\rho}{2}}E(\tau_2-\sigma)A^{\frac{r}{2}}g(\tau_2,Y(\tau_2))\varphi_m\right\|^2\,d\sigma$$

$$\leq C(s,r)\left\|A^{\frac{\eta}{2}}\right\|^2\left\|A^{\frac{r}{2}}g(\tau_2,Y(\tau_2))\right\|_{L_2^0}^2(\tau_2-\tau_1)^{\min(1,1+r-s)},$$

where $(\varphi_m)_{m\geq1}$ denotes an orthonormal basis of U_0. We also used that $1 - \rho = 1 - \max(0, s - r) = \min(1, 1 + r - s)$. Finally, by Assumption 2.20 we conclude

$$S_1 \leq C(r,s,A,g)\left(1 + \sup_{t\in[0,T]}\left\|Y(t)\right\|_{L^p(\Omega;\dot{H}^r)}\right)(\tau_2-\tau_1)^{\min(\frac{1}{2},\frac{1+r-s}{2})}.$$

This proves (2.44).

For S_2 we first recall from Proposition 2.9(iii) the fact that $\|B\Phi\|_{L_2^0} \leq \|B\|\|\Phi\|_{L_2^0}$ for all $B \in L(H)$ and $\Phi \in L_2^0$. Then we apply Lemma B.9(i) and get

$$S_2 \leq C(s, A) \left\| \left(\int_{\tau_1}^{\tau_2} (\tau_2 - \sigma)^{-s} \|g(\sigma, Y(\sigma)) - g(\tau_2, Y(\tau_2))\|_{L_2^0}^2 \, d\sigma \right)^{\frac{1}{2}} \right\|_{L^p(\Omega; \mathbb{R})}$$

$$= C(s, A) \left(\left\| \int_{\tau_1}^{\tau_2} (\tau_2 - \sigma)^{-s} \|g(\sigma, Y(\sigma)) - g(\tau_2, Y(\tau_2))\|_{L_2^0}^2 \, d\sigma \right\|_{L^{p/2}(\Omega; \mathbb{R})} \right)^{\frac{1}{2}}$$

$$\leq C(s, A) \left(\int_{\tau_1}^{\tau_2} (\tau_2 - \sigma)^{-s} \|g(\sigma, Y(\sigma)) - g(\tau_2, Y(\tau_2))\|_{L^p(\Omega; L_2^0)}^2 \, d\sigma \right)^{\frac{1}{2}}.$$

By Lemma 2.26 and the Hölder continuity of Y we arrive at

$$S_2 \leq C(s, A, g) \left(\int_{\tau_1}^{\tau_2} (\tau_2 - \sigma)^{-s+2\delta} \left(1 + \|Y(\sigma)\|_{L^p(\Omega; H)} \right)^2 \, d\sigma \right)^{\frac{1}{2}}$$

$$\leq \frac{C(s, A, g, C_\delta)}{\sqrt{1 + 2\delta - s}} \left(1 + \sup_{t \in [0, T]} \|Y(t)\|_{L^p(\Omega; H)} \right) (\tau_2 - \tau_1)^{\frac{1 + 2\delta - s}{2}}.$$

This shows (2.45). Combining the estimates for S_1 and S_2 yields (2.47) and, hence, (2.46) by using $(\tau_2 - \tau_1)^{2\delta - r} \leq T^{2\delta - r}$. □

Now we are well prepared for the proof of Theorem 2.27.

Proof (of Theorem 2.27). By taking norms in (2.20) we get, for $t \in [0, T]$,

$$\left(\mathbf{E} \left[\|X(t)\|_{r+1}^p \right] \right)^{\frac{1}{p}} = \left\| A^{\frac{r+1}{2}} X(t) \right\|_{L^p(\Omega; H)}$$

$$\leq \left\| A^{\frac{r+1}{2}} E(t) X_0 \right\|_{L^p(\Omega; H)}$$

$$+ \left\| A^{\frac{r+1}{2}} \int_0^t E(t - \sigma) f(\sigma, X(\sigma)) \, d\sigma \right\|_{L^p(\Omega; H)}$$

$$+ \left\| A^{\frac{r+1}{2}} \int_0^t E(t - \sigma) g(\sigma, X(\sigma)) \, dW(\sigma) \right\|_{L^p(\Omega; H)}$$

$$=: I + II + III.$$

The first term is well-known from deterministic theory and can be estimated by

$$\left\| A^{\frac{r+1}{2}} E(t) X_0 \right\|_{L^p(\Omega; H)} \leq \|X_0\|_{L^p(\Omega; \dot{H}^{1+r})} < \infty,$$

since $X_0 \colon \Omega \to \dot{H}^{r+1}$ by Assumption 2.17.

We recall that, by Theorem 2.25, the mild solution X is a predictable stochastic process which almost surely takes values in $\dot{H}^r$. In addition, X is δ-Hölder continuous for any $0 < \delta < \frac{1}{2}$ with respect to the norm $\| \cdot \|_{L^p(\Omega; H)}$. We choose $\delta := \frac{r+1}{4}$ so that $0 \leq \frac{r}{2} < \delta < \frac{1}{2}$. Hence, we can apply Lemmas 2.28 and 2.29 with $Y = X$.

For the second term we apply (2.43) with $\tau_1 = 0$, $\tau_2 = t$, $s = r + 1$ and $Y = X$. This yields

$$II \leq C\left(1 + \sup_{\sigma \in [0,T]} \|X(\sigma)\|_{L^p(\Omega;\dot{H}^r)}\right) < \infty.$$

For the last term we apply (2.46) with the same parameters as above:

$$III \leq C\left(1 + \sup_{\sigma \in [0,T]} \|X(\sigma)\|_{L^p(\Omega;\dot{H}^r)}\right) < \infty.$$

Note that $\sup_{\sigma \in [0,T]} \|X(\sigma)\|_{L^p(\Omega;\dot{H}^r)}$ is finite because of Theorem 2.25.   □

In order to motivate why we speak of optimal spatial regularity we conclude this section with the following example, where our result turns out to be sharp. Without loss of generality we restrict our discussion to the case $p = 2$. For $p > 2$ one may use the results on the optimal regularity of the stochastic convolution from [67] or [8].

Example 2.30 (continuation of Example 2.21). Let $H = U = L^2([0, 1]; \mathbb{R})$ be the space of all square integrable real-valued functions which are defined on the unit interval $[0, 1]$. Further, assume that $-A$ is the Laplacian with homogeneous Dirichlet boundary conditions. In this situation the eigenvalues and orthonormal eigenbasis of $-A$ are explicitly known to be

$$\lambda_n = n^2\pi^2 \quad \text{and} \quad e_n(y) = \sqrt{2}\sin(n\pi y) \quad \text{for all } n \geq 1, \ y \in [0, 1].$$

Consider the stochastic heat equation

$$dX(t) + AX(t)\,dt = dW(t), \quad \text{for } t \in [0, T],$$
$$X(0) = 0. \tag{2.48}$$

We choose W to be a Q-Wiener process on H, where the covariance operator $Q: H \to H$ is given by

$$Qe_1 = 0, \quad Qe_n = \frac{1}{n\log(n)^2}e_n \quad \text{for all } n \geq 2.$$

Then we have

$$\|\mathrm{Id}_H\|_{L^0_{2,r}}^2 = \sum_{n=2}^{\infty}\|A^{\frac{r}{2}}Q^{\frac{1}{2}}e_n\|^2 = \sum_{n=2}^{\infty}\lambda_n^r\frac{1}{n\log(n)^2} = \pi^{2r}\sum_{n=2}^{\infty}\frac{n^{2r}}{n\log(n)^2}.$$

Since this series converges only with $r = 0$, Assumption 2.20 is satisfied only for $r = 0$. Theorem 2.27 yields that the mild solution X to (2.14) takes values in $\dot{H}^1$. In the following we show that this result cannot be improved.

In this situation the mild formulation (2.20) reads

$$X(t) = \int_0^t E(t - \sigma) \, dW(\sigma).$$

Hence, by the Itô-isometry for the stochastic integral we have

$$
\begin{aligned}
\mathbf{E}\big[\big\|A^{\frac{1+r}{2}} X(t)\big\|^2\big] &= \int_0^t \big\|A^{\frac{1+r}{2}} E(t - \sigma)\big\|_{L_2^0}^2 \, d\sigma \\
&= \int_0^t \sum_{n=2}^{\infty} \lambda_n^{1+r} e^{-2\lambda_n(t-\sigma)} \frac{1}{n \log(n)^2} \, d\sigma \\
&= \frac{1}{2} \sum_{n=2}^{\infty} \lambda_n^r \big(1 - e^{-2\lambda_n t}\big) \frac{1}{n \log(n)^2} \\
&\geq \frac{1}{2} \pi^{2r} \big(1 - e^{-2\lambda_1 t}\big) \sum_{n=2}^{\infty} \frac{n^{2r}}{n \log(n)^2} = \infty \quad \text{for all } t > 0, \, r > 0.
\end{aligned}
$$

Thus, $X(t) \notin L^2(\Omega; \dot{H}^{1+r})$ for all $t > 0, r > 0$.

2.6 Temporal Regularity

This section is devoted to the temporal regularity of the mild solution. Our main result is summarized in the following theorem. For a result on the border case $s = r + 1$ we refer to Theorem 2.33 below.

Theorem 2.31. *Let $r \in [0, 1)$, $p \in [2, \infty)$ be given such that Assumptions 2.13–2.17 and Assumptions 2.19 and 2.20 are fulfilled. Then, for every $s \in [0, r + 1)$ the unique mild solution X to (2.14) is Hölder continuous with respect to the norm $\| \cdot \|_{L^p(\Omega; \dot{H}^s)}$ and satisfies*

$$
\sup_{\substack{t_1, t_2 \in [0,T] \\ t_1 \neq t_2}} \frac{\big(\mathbf{E}\big[\|X(t_1) - X(t_2)\|_s^p\big]\big)^{\frac{1}{p}}}{|t_1 - t_2|^{\min(\frac{1}{2}, \frac{1+r-s}{2})}} < \infty. \tag{2.49}
$$

Proof. Let $0 \leq t_1 < t_2 \leq T$ be arbitrary. By using the mild formulation (2.20) we get, for $s \in [0, r + 1)$,

$$
\begin{aligned}
&\big\|X(t_1) - X(t_2)\big\|_{L^p(\Omega; \dot{H}^s)} \\
&\leq \big\|(E(t_1) - E(t_2))X_0\big\|_{L^p(\Omega; \dot{H}^s)}
\end{aligned}
$$

$$+ \left\| \int_{t_1}^{t_2} E(t_2 - \sigma) f(\sigma, X(\sigma)) \, d\sigma \right\|_{L^p(\Omega; \dot{H}^s)}$$

$$+ \left\| \int_0^{t_1} \left(E(t_2 - \sigma) - E(t_1 - \sigma) \right) f(\sigma, X(\sigma)) \, d\sigma \right\|_{L^p(\Omega; \dot{H}^s)}$$

$$+ \left\| \int_{t_1}^{t_2} E(t_2 - \sigma) g(\sigma, X(\sigma)) \, dW(\sigma) \right\|_{L^p(\Omega; \dot{H}^s)}$$

$$+ \left\| \int_0^{t_1} \left(E(t_2 - \sigma) - E(t_1 - \sigma) \right) g(\sigma, X(\sigma)) \, dW(\sigma) \right\|_{L^p(\Omega; \dot{H}^s)}$$

$$=: T_1 + T_2 + T_3 + T_4 + T_5. \tag{2.50}$$

We estimate these five terms separately. For the term T_1 we get

$$T_1 = \left\| A^{\frac{s-r-1}{2}} (\mathrm{Id}_H - E(t_2 - t_1)) A^{\frac{r+1}{2}} E(t_1) X_0 \right\|_{L^p(\Omega; H)}$$

$$\leq C \left\| X_0 \right\|_{L^p(\Omega; \dot{H}^{1+r})} (t_2 - t_1)^{\frac{1+r-s}{2}},$$

where we used Lemma B.9(ii) and Assumption 2.17.

As in the proof of Theorem 2.27 we choose the Hölder exponent $\delta := \frac{r+1}{4}$ so that $\frac{r}{2} < \delta < \frac{1}{2}$ and we can apply Lemmas 2.28 and 2.29 with $Y = X$ by (2.27). The term T_2 then coincides with (2.43) and we have

$$T_2 \leq C \left(1 + \sup_{\sigma \in [0,T]} \| X(\sigma) \|_{L^p(\Omega; \dot{H}^r)} \right) (t_2 - t_1)^{\frac{1+r-s}{2}}.$$

For the third term we also apply Lemma B.9(ii) before we use (2.43):

$$T_3 = \left\| A^{\frac{s-r-1}{2}} \left(E(t_2 - t_1) - \mathrm{Id}_H \right) A^{\frac{r+1}{2}} \int_0^{t_1} E(t_1 - \sigma) f(\sigma, X(\sigma)) \, d\sigma \right\|_{L^p(\Omega; H)}$$

$$\leq C (t_2 - t_1)^{\frac{1+r-s}{2}} \left\| A^{\frac{r+1}{2}} \int_0^{t_1} E(t_1 - \sigma) f(\sigma, X(\sigma)) \, d\sigma \right\|_{L^p(\Omega; H)}$$

$$\leq C \left(1 + \sup_{\sigma \in [0,T]} \| X(\sigma) \|_{L^p(\Omega; \dot{H}^r)} \right) (t_2 - t_1)^{\frac{1+r-s}{2}}.$$

The fourth term is estimated analogously by using (2.46) instead of (2.43). We get

$$T_4 \leq C \left(1 + \sup_{\sigma \in [0,T]} \| X(\sigma) \|_{L^p(\Omega; \dot{H}^r)} \right) (t_2 - t_1)^{\min(\frac{1}{2}, \frac{1+r-s}{2})}.$$

Finally, for the last term we use Proposition 2.12 first. Since, for $0 \leq t_1 < t_2 \leq T$ fixed, the function $[0, t_1] \ni \sigma \mapsto A^{\frac{s}{2}} \left(E(t_2 - \sigma) - E(t_1 - \sigma) \right) g(\sigma, X(\sigma))$ is a predictable stochastic process Proposition 2.12 can be applied. Then, by using Lemma B.9(ii) with $\nu = \frac{1+r-s}{2}$ and Lemma 2.29 with $s = r + 1$ we get

$$T_5 \le C \left\| \left(\int_0^{t_1} \left\| A^{\frac{s-r-1}{2}} (E(t_2-t_1)-I) A^{\frac{r+1}{2}} E(t_1-\sigma) g(\sigma, X(\sigma)) \right\|_{L_2^0}^2 d\sigma \right)^{\frac{1}{2}} \right\|_{L^p(\Omega;\mathbb{R})}$$

$$\le C(t_2-t_1)^{\frac{1+r-s}{2}} \Bigg[\left\| \left(\int_0^{t_1} \left\| A^{\frac{r+1}{2}} E(t_1-\sigma) g(t_1, X(t_1)) \right\|_{L_2^0}^2 d\sigma \right)^{\frac{1}{2}} \right\|_{L^p(\Omega;\mathbb{R})}$$

$$+ \left\| \left(\int_0^{t_1} \left\| A^{\frac{r+1}{2}} E(t_1-\sigma) \big(g(\sigma, X(\sigma)) - g(t_1, X(t_1)) \big) \right\|_{L_2^0}^2 d\sigma \right)^{\frac{1}{2}} \right\|_{L^p(\Omega;\mathbb{R})} \Bigg]$$

$$\le C(t_2-t_1)^{\frac{1+r-s}{2}} \left(1 + \sup_{\sigma \in [0,T]} \| X(\sigma) \|_{L^p(\Omega;\dot{H}^r)} \right).$$

Altogether, this proves (2.49) and the Hölder continuity of X with respect to the norm $\| \cdot \|_{L^p(\Omega;\dot{H}^s)}$ for all $s \in [0, r+1)$. □

Example 2.32 (continuation of Example 2.24). We again consider the SODE

$$dX(t) = dW(t), \quad X(0) = 0, \tag{2.51}$$

with $H = U = \mathbb{R}$ and W a real-valued standard Wiener process, which in Example 2.24 is written as a semilinear SEEq on H. The solution is given by

$$X(t) = W(t), \quad t \in [0, T],$$

and it holds

$$\| W(t_1) - W(t_2) \|_{L^p(\Omega;H)} = \| W(t_1) - W(t_2) \|_{L^p(\Omega;\dot{H}^r)} = |t_1 - t_2|^{\frac{1}{2}}$$

for every $r \in \mathbb{R}$ since $H = \dot{H}^r = \mathbb{R}$.

Therefore, as long as a stochastic integral appears in the SEEq, one cannot expect to remove the upper bound $\frac{1}{2}$ of the Hölder exponent in (2.49).

The temporal regularity of X with respect to the norm $\| \cdot \|_{L^p(\Omega;\dot{H}^{r+1})}$ is more involved. For the case $r = 0$ we can prove the following result.

Theorem 2.33. *Let $p \in [2, \infty)$ be given such that Assumptions 2.13–2.17 are fulfilled with $r = 0$. Then the unique mild solution X to (2.14) is continuous with respect to the norm $\| \cdot \|_{L^p(\Omega;\dot{H}^1)}$.*

Proof. We show that

$$\lim_{\substack{t_2-t_1 \to 0 \\ t_1 < t_2}} \| X(t_2) - X(t_1) \|_{L^p(\Omega;\dot{H}^1)} = 0$$

with either t_1 or t_2 fixed. As already demonstrated in the proof of Lemma B.10 Lebesgue's dominated convergence theorem is our most important tool.

We consider again the terms T_i, $i = 1, \ldots, 5$, in (2.50) but now with $s = 1$.

For T_1 continuity follows immediately: For almost every $\omega \in \Omega$ we get that $X_0(\omega) \in \dot{H}^1$ by Assumption 2.17. Thus, for every fixed $\omega \in \Omega$ with this property we have

$$\lim_{t_2 - t_1 \to 0} \|(E(t_2) - E(t_1))A^{\frac{1}{2}}X_0(\omega)\| = 0$$

by the strong continuity of the semigroup, see Lemma B.3. We also have that

$$\|(E(t_2) - E(t_1))A^{\frac{1}{2}}X_0(\omega)\| \leq \|A^{\frac{1}{2}}X_0(\omega)\|,$$

where the latter is an element of $L^p(\Omega; \mathbb{R})$ as a function of $\omega \in \Omega$, which follows again from Assumption 2.17. Hence, Lebesgue's theorem is applicable and yields $\lim_{t_2 - t_1 \to 0} T_1 = 0$.

In order to treat the right and left limits simultaneously in the remaining terms, we compute the limits as $t_1 \to t_3$ and $t_2 \to t_3$ for fixed but arbitrary $t_3 \in [t_1, t_2]$.

In the case of T_2 we get

$$T_2 \leq \left\| A \int_{t_1}^{t_2} E(t_2 - \sigma)A^{-\frac{1}{2}}\big(f(\sigma, X(\sigma)) - f(t_2, X(t_2))\big)\, d\sigma \right\|_{L^p(\Omega;H)}$$

$$+ \left\| A \int_{t_1}^{t_2} E(t_2 - \sigma)A^{-\frac{1}{2}}\big(f(t_2, X(t_2)) - f(t_2, X(t_3))\big)\, d\sigma \right\|_{L^p(\Omega;H)} \qquad (2.52)$$

$$+ \left\| A \int_{t_1}^{t_2} E(t_2 - \sigma)A^{-\frac{1}{2}} f(t_3, X(t_3))\, d\sigma \right\|_{L^p(\Omega;H)}.$$

Because of (2.42), where we can choose $s = r + 1 = 1$ and $\delta = \frac{1}{4} > 0$, the limit $t_2 - t_1 \to 0$ of the first summand is 0. For the second summand in (2.52) we apply Lemma B.9(iv) with $\rho = 1$, and Assumption 2.14. Then we derive

$$\left\| A \int_{t_1}^{t_2} E(t_2 - \sigma)A^{-\frac{1}{2}}\big(f(t_2, X(t_2)) - f(t_3, X(t_3))\big)\, d\sigma \right\|_{L^p(\Omega;H)}$$

$$\leq C \left\| A^{-\frac{1}{2}}\big(f(t_2, X(t_2)) - f(t_3, X(t_3))\big) \right\|_{L^p(\Omega;H)}$$

and the limit $t_2 \to t_3$ of this term vanishes by Lemma 2.26 and (2.49) with $s = 0$.

For the last summand in (2.52) we again apply Lemma B.9(iv) with $\rho = 1$ and obtain, for almost every $\omega \in \Omega$,

$$\left\| A \int_{t_1}^{t_2} E(t_2 - \sigma)A^{-\frac{1}{2}} f(t_3, \omega, X(t_3, \omega))\, d\sigma \right\| \leq C \left\| A^{-\frac{1}{2}} f(t_3, \omega, X(t_3, \omega)) \right\|$$

and from Assumption 2.14 we obtain the following linear growth result on f

$$\left\| f(t_3, \omega, X(t_3, \omega)) \right\|_{-1} \leq \left\| f(t_3, \omega, X(t_3, \omega)) - f(0, \omega, 0) \right\|_{-1} + \left\| f(0, \omega, 0) \right\|_{-1}$$

$$\leq C\big(1 + \|X(t_3, \omega)\|\big)\big(1 + t_3^{\frac{1}{2}}\big) \qquad (2.53)$$

which shows that $\|f(t_3, \omega, X(t_3, \omega))\|_{-1} \in L^p(\Omega; \mathbb{R})$ for all $t_3 \in [0, T]$. By Lemma B.10(ii) it also holds that

$$\lim_{\substack{t_1 \to t_3 \\ t_2 \to t_3}} \left\| A \int_{t_1}^{t_2} E(t_2 - \sigma) A^{-\frac{1}{2}} f(t_3, \omega, X(t_3, \omega)) \, d\sigma \right\| = 0$$

for almost all $\omega \in \Omega$. Thus, Lebesgue's dominated convergence theorem yields that this term vanishes, which completes the proof for T_2.

Next, we take care of T_3, which is estimated by

$$T_3 \leq \left\| \int_0^{t_1} \big(E(t_2 - \sigma) - E(t_1 - \sigma) \big) \big(f(\sigma, X(\sigma)) - f(t_1, X(t_1)) \big) \, d\sigma \right\|_{L^p(\Omega; \dot{H}^1)}$$

$$+ \left\| \int_0^{t_1} \big(E(t_2 - \sigma) - E(t_1 - \sigma) \big) f(t_1, X(t_1)) \, d\sigma \right\|_{L^p(\Omega; \dot{H}^1)}.$$

$$(2.54)$$

For the first summand in (2.54) we get by Lemma B.9(i) and (ii)

$$\left\| \int_0^{t_1} \big(E(t_2 - \sigma) - E(t_1 - \sigma) \big) \big(f(\sigma, X(\sigma)) - f(t_1, X(t_1)) \big) \, d\sigma \right\|_{L^p(\Omega; \dot{H}^1)}$$

$$\leq \int_0^{t_1} \left\| A^{-\frac{\eta}{2}} \big(E(t_2 - t_1) - I \big) A^{\frac{1+\eta}{2}} E(t_1 - \sigma) \big(f(\sigma, X(\sigma)) \right.$$

$$\left. - f(t_1, X(t_1)) \big) \right\|_{L^p(\Omega; H)} \, d\sigma$$

$$\leq C(t_2 - t_1)^{\frac{\eta}{2}} \int_0^{t_1} (t_1 - \sigma)^{-\frac{2+\eta}{2}} \| f(\sigma, X(\sigma)) - f(t_1, X(t_1)) \|_{L^p(\Omega; \dot{H}^{-1})} \, d\sigma,$$

$$(2.55)$$

for some arbitrary $\eta \in (0, 1)$. We continue the estimate by applying Lemma 2.26 and the Hölder continuity of X with exponent $\frac{1}{2}$ with respect to the norm $\|\cdot\|_{L^p(\Omega; H)}$ as it was shown in (2.49). This gives

$$\left\| A \int_0^{t_1} \big(E(t_2 - \sigma) - E(t_1 - \sigma) \big) A^{-\frac{1}{2}} \big(f(\sigma, X(\sigma)) - f(t_1, X(t_1)) \big) \, d\sigma \right\|_{L^p(\Omega; H)}$$

$$\leq C(t_2 - t_1)^{\frac{\eta}{2}} \int_0^{t_1} (t_1 - \sigma)^{-\frac{2+\eta-1}{2}} \, d\sigma \Big(1 + \sup_{t \in [0,T]} \| X(t) \|_{L^p(\Omega; H)} \Big)$$

$$= C \frac{2}{1 - \eta} t_1^{\frac{1-\eta}{2}} (t_2 - t_1)^{\frac{\eta}{2}} \Big(1 + \sup_{t \in [0,T]} \| X(t) \|_{L^p(\Omega; H)} \Big).$$

Therefore, in the limit $t_2 - t_1 \to 0$ this term goes to zero.

For the second summand in (2.54) we apply Lemma B.9(*iv*) with $\rho = 1$ and get

$$\left\| \int_0^{t_1} \left(E(t_2 - \sigma) - E(t_1 - \sigma) \right) f(t_1, X(t_1)) \, d\sigma \right\|_{L^p(\Omega;\dot{H}^1)}$$

$$= \left\| A \int_0^{t_1} E(t_1 - \sigma) \left(E(t_2 - t_1) - \mathrm{Id}_H \right) A^{-\frac{1}{2}} f(t_1, X(t_1)) \, d\sigma \right\|_{L^p(\Omega;H)}$$

$$\leq C \left\| \left(E(t_2 - t_1) - \mathrm{Id}_H \right) A^{-\frac{1}{2}} f(t_1, X(t_1)) \right\|_{L^p(\Omega;H)}$$

$$\leq C \left\| \left(E(t_2 - t_1) - \mathrm{Id}_H \right) A^{-\frac{1}{2}} \left(f(t_1, X(t_1)) - f(t_3, X(t_3)) \right) \right\|_{L^p(\Omega;H)}$$

$$+ C \left\| \left(E(t_2 - t_1) - \mathrm{Id}_H \right) A^{-\frac{1}{2}} f(t_3, X(t_3)) \right\|_{L^p(\Omega;H)}.$$

By Lemma 2.26 and (2.49) it holds true that

$$\left\| \left(E(t_2 - t_1) - \mathrm{Id}_H \right) A^{-\frac{1}{2}} \left(f(t_1, X(t_1)) - f(t_3, X(t_3)) \right) \right\|_{L^p(\Omega;H)}$$

$$\leq C \left(1 + \sup_{t \in [0,T]} \| X(t) \|_{L^p(\Omega;H)} \right) |t_1 - t_3|^{\frac{1}{2}}.$$

Hence, this term vanishes in the limit $t_1 \to t_3$.

Therefore, the proof for T_3 is complete, if we can show that

$$\lim_{\substack{t_1 \to t_3 \\ t_2 \to t_3}} \left\| \left(E(t_2 - t_1) - \mathrm{Id}_H \right) A^{-\frac{1}{2}} f(t_3, X(t_3)) \right\|_{L^p(\Omega;H)} = 0.$$

This is true by an application of Lebesgue's dominated convergence theorem. In order to apply this theorem, we obtain a dominating function for almost every $\omega \in \Omega$ by

$$\left\| \left(E(t_2 - t_1) - \mathrm{Id}_H \right) A^{-\frac{1}{2}} f(t_3, \omega, X(t_3, \omega)) \right\| \leq C \left(1 + \| X(t_3, \omega) \| \right),$$

where we applied the same linear growth bound on f as in (2.53). Further, Lemma B.3 yields

$$\lim_{\substack{t_1 \to t_3 \\ t_2 \to t_3}} \left\| \left(E(t_2 - t_1) - \mathrm{Id}_H \right) A^{-\frac{1}{2}} f(t_3, \omega, X(t_3, \omega)) \right\| = 0$$

for almost every $\omega \in \Omega$. Altogether, this shows $\lim_{\substack{t_1 \to t_3 \\ t_2 \to t_3}} T_3 = 0$.

For T_4, one has to use Proposition 2.12, which yields

$$
T_4 \leq C \left\| \left(\int_{t_1}^{t_2} \left\| A^{\frac{1}{2}} E(t_2 - \sigma) g(\sigma, X(\sigma)) \right\|_{L_2^0}^2 d\sigma \right)^{\frac{1}{2}} \right\|_{L^p(\Omega;\mathbb{R})}
$$

$$
\leq C \left\| \left(\int_{t_1}^{t_2} \left\| A^{\frac{1}{2}} E(t_2 - \sigma) \big(g(\sigma, X(\sigma)) - g(t_2, X(t_2)) \big) \right\|_{L_2^0}^2 d\sigma \right)^{\frac{1}{2}} \right\|_{L^p(\Omega;\mathbb{R})}
$$

$$
+ C \left\| \left(\int_{t_1}^{t_2} \left\| A^{\frac{1}{2}} E(t_2 - \sigma) \big(g(t_2, X(t_2)) - g(t_3, X(t_3)) \big) \right\|_{L_2^0}^2 d\sigma \right)^{\frac{1}{2}} \right\|_{L^p(\Omega;\mathbb{R})}
$$

$$
+ C \left\| \left(\int_{t_1}^{t_2} \left\| A^{\frac{1}{2}} E(t_2 - \sigma) g(t_3, X(t_3)) \right\|_{L_2^0}^2 d\sigma \right)^{\frac{1}{2}} \right\|_{L^p(\Omega;\mathbb{R})}.
$$

$$(2.56)$$

The limit $t_2 - t_1 \to 0$ of the first summand is 0 because of (2.45), where we again choose $s = 1$ and $\delta = \frac{1}{4} > \frac{r}{2} = 0$. As before, we discuss the simultaneous limits $t_1 \to t_3$ and $t_2 \to t_3$ for the remaining summands in (2.56).

By Lemma B.9(*iii*) it holds for the second summand in (2.56) that

$$
\left\| \left(\int_{t_1}^{t_2} \left\| A^{\frac{1}{2}} E(t_2 - \sigma) \big(g(t_2, X(t_2)) - g(t_3, X(t_3)) \big) \right\|_{L_2^0}^2 d\sigma \right)^{\frac{1}{2}} \right\|_{L^p(\Omega;\mathbb{R})}
$$

$$
= \left\| \left(\sum_{m=1}^{\infty} \int_{t_1}^{t_2} \left\| A^{\frac{1}{2}} E(t_2 - \sigma) \big(g(t_2, X(t_2)) - g(t_3, X(t_3)) \big) \varphi_m \right\|^2 d\sigma \right)^{\frac{1}{2}} \right\|_{L^p(\Omega;\mathbb{R})}
$$

$$
\leq C \left\| g(t_2, X(t_2)) - g(t_3, X(t_3)) \right\|_{L^p(\Omega;L_2^0)},
$$

$$(2.57)$$

where $(\varphi_m)_{m \geq 1}$ is an arbitrary orthonormal basis of U_0. Consequently, this term also vanishes as $t_2 \to t_3$ by Lemma 2.26.

Next we come to the third summand in (2.56). By Lemma B.9(*iii*) with $\rho = 1$ we obtain for almost every $\omega \in \Omega$

$$
\left(\int_{t_1}^{t_2} \left\| A^{\frac{1}{2}} E(t_2 - \sigma) g(t_3, \omega, X(t_3, \omega)) \right\|_{L_2^0}^2 d\sigma \right)^{\frac{1}{2}}
$$

$$
= \left(\sum_{m=1}^{\infty} \int_{t_1}^{t_2} \left\| A^{\frac{1}{2}} E(t_2 - \sigma) g(t_3, \omega, X(t_3, \omega)) \varphi_m \right\|^2 d\sigma \right)^{\frac{1}{2}}
$$

$$
\leq C \left\| g(t_3, \omega, X(t_3, \omega)) \right\|_{L_2^0},
$$

where the last term belongs to $L^p(\Omega; \mathbb{R})$ by Lemma 2.26.

In order to apply Lebesgue's Theorem it remains to discuss the ω-wise limit. For this Lemma B.10(*i*) yields

$$\lim_{\substack{t_1 \to t_3 \\ t_2 \to t_3}} \int_{t_1}^{t_2} \left\| A^{\frac{1}{2}} E(t_2 - \sigma) g(t_3, \omega, X(t_3, \omega)) \right\|_{L_2^0}^2 d\sigma$$

$$= \sum_{m=1}^{\infty} \lim_{\substack{t_1 \to t_3 \\ t_2 \to t_3}} \int_{t_1}^{t_2} \left\| A^{\frac{1}{2}} E(t_2 - \sigma) g(t_3, \omega, X(t_3, \omega)) \varphi_m \right\|^2 d\sigma = 0.$$

In fact, the interchanging of summation and taking the limit is justified by a further application of Lebesgue's Theorem. Altogether, this proves the desired result for T_4.

The estimate of T_5 works similarly as for T_3. We apply Proposition 2.12 and get

$$T_5 \leq C \left\| \left(\int_0^{t_1} \left\| A^{\frac{1}{2}} (E(t_2 - \sigma) - E(t_1 - \sigma)) g(t_1, X(t_1)) \right\|_{L_2^0}^2 d\sigma \right)^{\frac{1}{2}} \right\|_{L^p(\Omega;\mathbb{R})}$$

$$+ C \left\| \left(\int_0^{t_1} \left\| A^{\frac{1}{2}} (E(t_2 - \sigma) - E(t_1 - \sigma)) \right. \right. \right. \tag{2.58}$$

$$\left. \left. \left. \times \left(g(\sigma, X(\sigma)) - g(t_1, X(t_1)) \right) \right\|_{L_2^0}^2 d\sigma \right)^{\frac{1}{2}} \right\|_{L^p(\Omega;\mathbb{R})}.$$

By using a similar technique as for (2.55) the second summand in (2.58) is estimated by

$$\left\| \left(\int_0^{t_1} \left\| A^{\frac{1}{2}} (E(t_2 - \sigma) - E(t_1 - \sigma)) (g(\sigma, X(\sigma)) - g(t_1, X(t_1))) \right\|_{L_2^0}^2 d\sigma \right)^{\frac{1}{2}} \right\|_{L^p(\Omega;\mathbb{R})}^2$$

$$= \left\| \int_0^{t_1} \left\| A^{\frac{1}{2}} (E(t_2 - \sigma) - E(t_1 - \sigma)) (g(\sigma, X(\sigma)) - g(t_1, X(t_1))) \right\|_{L_2^0}^2 d\sigma \right\|_{L^{p/2}(\Omega;\mathbb{R})}$$

$$\leq C(t_2 - t_1)^\eta \left\| \int_0^{t_1} (t_1 - \sigma)^{-1-\eta} \left\| g(\sigma, X(\sigma)) - g(t_1, X(t_1)) \right\|_{L_2^0}^2 d\sigma \right\|_{L^{p/2}(\Omega;\mathbb{R})}$$

$$\leq C(t_2 - t_1)^\eta \int_0^{t_1} (t_1 - \sigma)^{-1-\eta} \left\| g(\sigma, X(\sigma)) - g(t_1, X(t_1)) \right\|_{L^p(\Omega;L_2^0)}^2 d\sigma.$$

For the first inequality we applied Lemma B.9(i) and (ii) with an arbitrary parameter $\eta \in (0, 1)$. Next we use Lemma 2.26 and continue as in (2.55). It follows that the summand vanishes in the limit $t_2 - t_1 \to 0$.

For the first summand in (2.58) it holds that

$$\left\| \left(\int_0^{t_1} \left\| A^{\frac{1}{2}} (E(t_2 - \sigma) - E(t_1 - \sigma)) g(t_1, X(t_1)) \right\|_{L_2^0}^2 d\sigma \right)^{\frac{1}{2}} \right\|_{L^p(\Omega;\mathbb{R})}$$

$$= \left\| \left(\sum_{m=1}^{\infty} \int_0^{t_1} \left\| A^{\frac{1}{2}} E(t_1 - \sigma) (E(t_2 - t_1) - \mathrm{Id}_H) g(t_1, X(t_1)) \varphi_m \right\|^2 d\sigma \right)^{\frac{1}{2}} \right\|_{L^p(\Omega;\mathbb{R})}$$

$$\leq C \left\| (E(t_2 - t_1) - \mathrm{Id}_H) g(t_1, X(t_1)) \right\|_{L^p(\Omega;L_2^0)}$$

$$\leq C \big\| g(t_1, X(t_1)) - g(t_3, X(t_3)) \big\|_{L^p(\Omega; L_2^0)}$$
$$+ \big\| \big(E(t_2 - t_1) - \mathrm{Id}_H \big) g(t_3, X(t_3)) \big\|_{L^p(\Omega; L_2^0)},$$

where we used Lemma B.9(iii). By Lemma 2.26 and (2.49) it holds that

$$\lim_{t_1 \to t_3} \big\| g(t_1, X(t_1)) - g(t_3, X(t_3)) \big\|_{L^p(\Omega; L_2^0)} = 0.$$

Since, as above, Lebesgue's dominated convergence theorem yields that

$$\lim_{\substack{t_1 \to t_3 \\ t_2 \to t_3}} \big\| \big(E(t_2 - t_1) - \mathrm{Id}_H \big) g(t_3, X(t_3)) \big\|_{L^p(\Omega; L_2^0)} = 0$$

the proof for T_5 is complete. This also completes the proof of the theorem. $\qquad \square$

Remark 2.34. If one wants to extend the result of Theorem 2.33 under the additional Assumptions 2.19 and 2.20 to general $r \in [0, 1)$ it is not hard to adapt the given arguments for all terms T_i, $i \in \{1, 3, 5\}$. For the other two terms, however, the situation is more delicate.

For T_4, this becomes apparent in the discussion of (2.57), which for $r \in (0, 1)$ is equal to

$$\left\| \left(\int_{t_1}^{t_2} \Big\| A^{\frac{1+r}{2}} E(t_2 - \sigma) \big(g(t_2, X(t_2)) - g(t_3, X(t_3)) \big) \Big\|_{L_2^0}^2 d\sigma \right)^{\frac{1}{2}} \right\|_{L^p(\Omega; \mathbb{R})}^2$$
$$\leq C \big\| A^{\frac{r}{2}} \big(g(t_2, X(t_2)) - g(t_3, X(t_3)) \big) \big\|_{L^p(\Omega; L_2^0)}.$$

Unlike the case $r = 0$ we do not want to make the restrictive assumption that for some $r > 0$ and all $t \in [0, T]$, $\omega \in \Omega$, the mapping $x \mapsto A^{\frac{r}{2}} g(t, x)$ is globally Lipschitz continuous on H^r. Therefore, we cannot directly conclude that this term vanishes in the limit $t_2 \to t_3$.

However, for $p \in (2, \infty]$, it is enough to assume that the mapping $(t, x) \mapsto A^{\frac{r}{2}} g(t, \omega, x)$ is continuous for all $\omega \in \Omega$. In order to show this we use a generalized version of Lebesgue's dominated convergence theorem (see Theorem C.1), which allows a t_2-dependent family of dominating functions.

In fact, by the linear growth condition in Assumption (2.22) we obtain, for almost all $\omega \in \Omega$,

$$\big\| g(t_2, \omega, X(t_2, \omega)) - g(t_3, \omega, X(t_3, \omega)) \big\|_{L_{2,r}^0} \leq C \big(1 + \|X(t_2, \omega)\|_r + \|X(t_3, \omega)\|_r \big),$$

where, by (2.49), the family of dominating functions converges:

$$\big(1 + \|X(t_2)\|_r + \|X(t_3)\|_r \big) \to \big(1 + 2\|X(t_3)\|_r \big) \quad \text{in } L^p(\Omega, \mathbb{R}) \text{ as } t_2 \to t_3.$$

Further, by (2.49) with $p \in (2, \infty)$ Kolmogorov's continuity theorem [18, Th. 3.3] yields that there exists a continuous version of the process $t \mapsto A^{\frac{r}{2}} X(t)$. Hence, under the additional assumption that $(t, x) \mapsto A^{\frac{r}{2}} g(t, \omega, x)$ is continuous for all $\omega \in \Omega$ we obtain

$$\lim_{t_2 \to t_3} \left\| A^{\frac{r}{2}} \big(g(t_2, \omega, X(t_2, \omega)) - g(t_3, \omega, X(t_3, \omega)) \big) \right\|_{L_2^0} = 0,$$

for almost all $\omega \in \Omega$. By Theorem C.1 we conclude that the unique mild solution X to (2.14) is continuous with respect to $\| \cdot \|_{L^p(\Omega; \dot{H}^{r+1})}$.

The case $p = 2$ remains as an open problem.

2.7 Some Generalizations

In this section we collect some possible extensions of our results to more general situations.

The first generalization is concerned with the operator A. As in [41] our proof of the existence and uniqueness theorem does not make use of the assumption that the inverse of A is compact. In particular, Theorem 2.25 remains valid under the more general condition that $-A$ is the generator of an analytic semigroup.

The same is true for the regularity results if we consider only the deterministic case, that is $g \equiv 0$. This is true since Lemma B.9(iv) also holds for analytic semigroups.

However, we could not find an answer to the question if Lemma B.9(iii) also remains valid for analytic semigroups. If this would be true, then our optimal spatial regularity results immediately carry over to analytic semigroups.

Secondly, we are concerned with Assumptions 2.14 and 2.16. In particular, in Chap. 6 we would like to treat inhomogeneities of the form $f(t) = W(t)$. In this case, f is a stochastic process but there does *not* exist a constant C such that ω-wise Hölder continuity (2.16) holds with exponent $\delta = \frac{1}{2}$. Consequently, Assumption 2.14 is not fulfilled.

However, we mainly use Assumptions 2.14 and 2.16 for the proof of Lemma 2.26. Since the assertion of Lemma 2.26 is also valid for $f(t) = W(t)$, the proof of Theorems 2.25, 2.27 and 2.31 also work with $f(t) = W(t)$.

Hence, for these Theorems, instead of Assumptions 2.14 and 2.16 we also could have assumed that the statement of Lemma 2.26 is true. Only the proof of Theorem 2.33 makes directly use of the pathwise assumptions.

Finally, we note that Assumptions 2.14 and 2.19 can be replaced by a suitable global Lipschitz condition on f as a mapping from $[0, T] \times \Omega \times H \to H$.

Chapter 3
Optimal Strong Error Estimates for Galerkin Finite Element Methods ·

This chapter contains our analysis of the strong error of convergence for Galerkin finite element approximations of stochastic evolution equations and is a slightly modified version of [49]. Our two main results in Sects. 3.4 and 3.6 are concerned with the error of the spatially semidiscrete approximation and of the spatio-temporal discretization of the mild solution.

The sections around the main results give a brief introduction to Galerkin finite element methods and prepare the proofs of the main results by providing sharp error estimates for the corresponding deterministic linear problem associated to our stochastic evolution equation.

Let us remark that for our error analysis of the deterministic linear problem we use the results from [66, Chaps. 2, 3 and 7] which are stated under the assumption that $-A$ is the Laplace operator with homogeneous Dirichlet boundary conditions. But all proofs and techniques also hold in our more general framework of a self-adjoint, positive definite operator A with compact inverse.

3.1 Preliminaries

Our aim is to study numerical methods which are used to approximate the mild solution $X: [0, T] \times \Omega \to H$ to the semilinear stochastic evolution equation

$$dX(t) + [AX(t) + f(t, X(t))] \, dt = g(t, X(t)) \, dW(t), \text{ for } 0 \le t \le T,$$
$$X(0) = X_0. \tag{3.1}$$

For given $p \in [2, \infty)$ and $r \in [0, 1)$ we work in the same setting and under the same assumptions as in Sect. 2.3. Thus, Theorem 2.25 ensures the existence of a unique mild solution which is given by

R. Kruse, *Strong and Weak Approximation of Semilinear Stochastic Evolution Equations*,
Lecture Notes in Mathematics 2093, DOI 10.1007/978-3-319-02231-4_3,
© Springer International Publishing Switzerland 2014

$$X(t) = E(t)X_0 - \int_0^t E(t - \sigma) f(\sigma, X(\sigma)) \, d\sigma$$
$$+ \int_0^t E(t - \sigma) g(\sigma, X(\sigma)) \, dW(\sigma) \quad \textbf{P}\text{-a.s.}$$
(3.2)

Further, by Theorems 2.27 and 2.31, we have

$$\sup_{t \in [0,T]} \mathbf{E}\left[\|X(t)\|_s^p\right] < \infty, \quad \text{for all } s \in [0, r+1], \tag{3.3}$$

and there exists a constant C such that

$$\left(\mathbf{E}\left[\|X(t_1) - X(t_2)\|_s^p\right]\right)^{\frac{1}{p}} \le C |t_1 - t_2|^{\frac{1}{2}} \tag{3.4}$$

for all $t_1, t_2 \in [0, T]$, $s \in [0, r]$.

For reasons which we explain in Sect. 3.4 it is convenient to replace the linear growth condition on f in Assumption 2.19 by a global Lipschitz conditions which is comparable to [20, Assumption 2]. As above let $r \in [0, 1)$.

Assumption 3.1. The mapping $f : [0, T] \times \Omega \times \dot{H}^r \to \dot{H}^{-1+r}$, $(t, \omega, h) \mapsto f(t, \omega, h)$ is $\mathscr{P}_T \times \mathscr{B}(\dot{H}^r)/\mathscr{B}(\dot{H}^{-1+r})$-measurable.

Further, there exists a constant $C > 0$ such that $\|f(0, \omega, 0)\|_{-1+r} \le C$ for all $\omega \in \Omega$, and

$$\|f(t, \omega, h_1) - f(t, \omega, h_2)\|_{-1+r} \le C \|h_1 - h_2\|_r \tag{3.5}$$

for all $h_1, h_2 \in \dot{H}^r$, $\omega \in \Omega$, $t \in [0, T]$. Also, there exists a constant $C > 0$ with

$$\|f(t_1, \omega, h) - f(t_2, \omega, h)\|_{-1+r} \le C(1 + \|h\|_r)(t_2 - t_1)^{\frac{1}{2}} \tag{3.6}$$

for all $h \in \dot{H}^r$, $0 \le t_1 < t_2 \le T$, $\omega \in \Omega$.

Note that instead of Assumptions 2.14 and 3.1 it is also sufficient to assume that f satisfies a global Lipschitz condition on H.

From Assumption 3.1 we obtain an analogue to Lemma 2.26.

Lemma 3.2. *Let Assumption 3.1 hold with* $r \in [0, 1)$ *and let* X *denote the mild solution to (3.1). Then there exists a constant* $C > 0$ *such that*

$$\left\|f(\tau_1, X(\tau_1)) - f(\tau_2, X(\tau_2))\right\|_{L^p(\Omega; \dot{H}^{-1+r})} \le C\left(1 + \|X(\tau_1)\|_{L^p(\Omega; \dot{H}^r)}\right)|\tau_1 - \tau_2|^{\frac{1}{2}}$$

for all $\tau_1, \tau_2 \in [0, T]$.

In particular, we have

$$\left\|f(\tau, X(\tau))\right\|_{L^p(\Omega; \dot{H}^{-1+r})} \le C\left(1 + \|X(\tau)\|_{L^p(\Omega; \dot{H}^r)}\right). \tag{3.7}$$

for all $\tau \in [0, T]$.

Proof. The proof works in the same way as the proof of Lemma 2.26. We have

$$\|f(\tau_1, X(\tau_1)) - f(\tau_2, X(\tau_2))\|_{L^p(\Omega;\dot{H}^{-1+r})}$$
$$\leq \|f(\tau_1, X(\tau_1)) - f(\tau_2, X(\tau_1))\|_{L^p(\Omega;\dot{H}^{-1+r})}$$
$$+ \|f(\tau_2, X(\tau_1)) - f(\tau_2, X(\tau_2))\|_{L^p(\Omega;\dot{H}^{-1+r})}.$$

For the first summand it holds by (3.6) that

$$\|f(\tau_1, X(\tau_1)) - f(\tau_2, X(\tau_1))\|_{L^p(\Omega;\dot{H}^{-1+r})}$$
$$= \left(\int_\Omega \|f(\tau_1, \omega, X(\tau_1, \omega)) - f(\tau_2, \omega, X(\tau_1, \omega))\|_{-1+r}^p \, d\mathbf{P}(\omega) \right)^{\frac{1}{p}}$$
$$\leq C |\tau_1 - \tau_2|^{\frac{1}{2}} \left(\int_\Omega \left(1 + \|X(\tau_1, \omega)\|_r\right)^p \, d\mathbf{P}(\omega) \right)^{\frac{1}{p}}$$
$$\leq C \left(1 + \|X(\tau_1)\|_{L^p(\Omega;\dot{H}^r)}\right) |\tau_1 - \tau_2|^{\frac{1}{2}}.$$

The second summand is estimated similarly by making use of (3.5). We obtain

$$\|f(\tau_2, X(\tau_1)) - f(\tau_2, X(\tau_2))\|_{L^p(\Omega;\dot{H}^{-1+r})}$$
$$= \left(\int_\Omega \|f(\tau_2, \omega, X(\tau_1, \omega)) - f(\tau_2, \omega, X(\tau_2, \omega))\|_{-1+r}^p \, d\mathbf{P}(\omega) \right)^{\frac{1}{p}}$$
$$\leq C \left(\int_\Omega \|X(\tau_1, \omega) - X(\tau_2, \omega)\|_r^p \, d\mathbf{P}(\omega) \right)^{\frac{1}{p}}$$
$$= C \|X(\tau_1) - X(\tau_2)\|_{L^p(\Omega;\dot{H}^r)} \leq C |\tau_1 - \tau_2|^{\frac{1}{2}},$$

where we also applied (3.4).

By Assumption 3.1 we further have $\|f(0,0)\|_{L^p(\Omega;\dot{H}^{-1+r})} < \infty$. Therefore,

$$\|f(\tau, X(\tau))\|_{L^p(\Omega;\dot{H}^{-1+r})} \leq \|f(\tau, X(\tau)) - f(0,0)\|_{L^p(\Omega;\dot{H}^{-1+r})}$$
$$+ \|f(0,0)\|_{L^p(\Omega;\dot{H}^{-1+r})}$$
$$\leq C \left(1 + \|X(\tau)\|_{L^p(\Omega;\dot{H}^r)}\right),$$

where the generic constant C also depends on T and $\|f(0,0)\|_{L^p(\Omega;\dot{H}^{-1+r})}$ but is independent of X. □

We close this section with the remark that, as already noted in Sect. 2.7, it is sufficient to replace Assumptions 2.14 and 3.1 by the assertion of Lemma 3.2. Then, all results of this chapter remain valid.

3.2   A Brief Introduction to Galerkin Finite Element Methods

In this section we first review the *Galerkin finite element methods* used for the discretization of the Hilbert space H. Following [66, Chaps. 2 and 3] we recall the definition of several discrete operators which are connected to a sequence of finite dimensional subspaces of $\dot{H}^1$. We close this section with two more concrete examples of spatial discretization schemes, namely the standard finite element method and a spectral Galerkin method.

First we recall from Appendix B.2 the notion of the Hilbert spaces $\dot{H}^s = \mathrm{dom}(A^{\frac{s}{2}})$. Let $(S_h)_{h \in (0,1]}$ be a sequence of finite dimensional subspaces of $\dot{H}^1$ and denote by $R_h \colon \dot{H}^1 \to S_h$ the orthogonal projector (or *Ritz projector*) onto S_h with respect to the inner product $(\cdot, \cdot)_1 = (A^{\frac{1}{2}} \cdot, A^{\frac{1}{2}} \cdot)$ in $\dot{H}^1$. Thus, we have

$$(R_h x, y_h)_1 = (x, y_h)_1 \text{ for all } x \in \dot{H}^1, \ y_h \in S_h.$$

Throughout this chapter we assume that the spaces $(S_h)_{h \in (0,1]}$ satisfy the following approximation property. Below we present two examples of A, H and spaces $(S_h)_{h \in (0,1]}$ which fulfill this assumption.

Assumption 3.3. Let a sequence $(S_h)_{h \in (0,1]}$ of finite dimensional subspaces of $\dot{H}^1$ be given such that there exists a constant C with

$$\|R_h x - x\| \le Ch^s \|x\|_s \text{ for all } x \in \dot{H}^s, \ s \in \{1, 2\}, \ h \in (0, 1]. \qquad (3.8)$$

Remark 3.4. Following [53, Chap. 5.2] or [66, Chap. 1] let us consider the linear (elliptic) problem to find $x \in \mathrm{dom}(A) = \dot{H}^2$ such that $Ax = z$ holds for a given $z \in H$. The *weak* or *variational* formulation of this problem is: Find $x \in \dot{H}^1$ which satisfies

$$(x, y)_1 = (z, y) \quad \text{for all } y \in \dot{H}^1. \qquad (3.9)$$

For a given sequence of finite dimensional subspaces $(S_h)_{h \in (0,1]}$ the Galerkin approximation $x_h \in S_h$ of the weak solution x is given by the relationship

$$(x_h, y_h)_1 = (z, y_h) \quad \text{for all } y_h \in S_h. \qquad (3.10)$$

Note that by the representation theorem $x \in \dot{H}^1$ and $x_h \in S_h$ are uniquely determined by (3.9) and (3.10). By the definition of the Ritz projector and since (3.9) holds for all $y_h \in S_h$ we get

$$(R_h x, y_h)_1 = (x, y_h)_1 = (x_h, y_h)_1 \quad \text{for all } y_h \in S_h.$$

This yields $R_h x = x_h$. Hence, the Ritz projection $R_h x$ coincides with the Galerkin approximation of the solution x to the elliptic problem. Hence, Assumption 3.3 is a statement about the order of convergence of the sequence $(x_h)_{h \in (0,1]}$ to x.

Next, we introduce the mapping $A_h : S_h \rightarrow S_h$, which is a discrete version of the operator A. For $x_h \in S_h$ we define $A_h x_h$ to be the unique element in S_h which satisfies the relationship

$$(x_h, y_h)_1 = (A_h x_h, y_h) \quad \text{for all } y_h \in S_h. \tag{3.11}$$

Since

$$(A_h x_h, y_h) = (x_h, y_h)_1 = (x_h, A_h y_h) \quad \text{for all } x_h, y_h \in S_h,$$

as well as

$$(A_h x_h, x_h) = (x_h, x_h)_1 = \|x_h\|_1^2 > 0 \quad \text{for all } x_h \in S_h, \ x_h \neq 0,$$

the operator A_h is self-adjoint and positive definite on S_h. Hence, $-A_h$ is the generator of an analytic semigroup of contractions on S_h, which is denoted by $E_h(t) := e^{-A_h t}$. Let $\rho \geq 0$. Similar to [66, Lemma 3.9] one shows the smoothing property

$$\left\| A_h^\rho E_h(t) y_h \right\| \leq C t^{-\rho} \|y_h\| \quad \text{for all } t > 0, \tag{3.12}$$

where $C = C(\rho)$ is independent of $h \in (0, 1]$. Additionally, by the definition of A_h, it holds true that

$$\left\| A_h^{\frac{1}{2}} y_h \right\|^2 = (y_h, y_h)_1 = \left\| A^{\frac{1}{2}} y_h \right\|^2 = \|y_h\|_1^2 \quad \text{for all } y_h \in S_h \subset \dot{H}^1. \tag{3.13}$$

Finally, let $P_h : \dot{H}^{-1} \rightarrow S_h$ be the generalized orthogonal projector onto S_h (see also [11]) defined by

$$(P_h x, y_h) = \langle x, y_h \rangle \quad \text{for all } x \in \dot{H}^{-1}, \ y_h \in S_h,$$

where $\langle \cdot, \cdot \rangle = \left(A^{-1} \cdot, \cdot \right)_1$ denotes the duality pairing between $\dot{H}^{-1}$ and $\dot{H}^1$ (see also Theorem B.8). By the representation theorem, P_h is well-defined and, when restricted to H, coincides with the usual orthogonal projector onto S_h with respect to the H-inner product.

These operators are related as follows (see [66, (2.16)]):

$$A_h^{-1} P_h x = R_h A^{-1} x \quad \text{for all } x \in \dot{H}^{-1}, \tag{3.14}$$

since

$$(R_h A^{-1} x, y_h)_1 = (A^{-1} x, y_h)_1 = \langle x, y_h \rangle = (P_h x, y_h) = (A_h^{-1} P_h x, y_h)_1$$

for all $x \in \dot{H}^{-1}$, $y_h \in S_h$. Furthermore, by Larsson [52, (3.7)] it holds that

$$
\begin{aligned}
\|A_h^{-\frac{1}{2}} P_h x\| &= \sup_{z_h \in S_h} \frac{|(A_h^{-\frac{1}{2}} P_h x, z_h)|}{\|z_h\|} = \sup_{z_h \in S_h} \frac{|(P_h x, A_h^{-\frac{1}{2}} z_h)|}{\|z_h\|} \\
&= \sup_{z_h' \in S_h} \frac{|\langle x, z_h' \rangle|}{\|A_h^{\frac{1}{2}} z_h'\|} \leq \sup_{z_h' \in S_h} \frac{\|x\|_{-1} \|z_h'\|_1}{\|A_h^{\frac{1}{2}} z_h'\|} = \|x\|_{-1},
\end{aligned}
\tag{3.15}
$$

for all $x \in \dot{H}^{-1}$, where the last equality is due to (3.13). Having this established we also prove the following consequence of (3.12):

$$
\|E_h(t) P_h x\| = \|A_h^{\frac{1}{2}} E_h(t) A_h^{-\frac{1}{2}} P_h x\| \leq C t^{-\frac{1}{2}} \|A_h^{-\frac{1}{2}} P_h x\| \leq C t^{-\frac{1}{2}} \|x\|_{-1} \tag{3.16}
$$

for all $x \in \dot{H}^{-1}$, $t > 0$ and $h \in (0, 1]$.

The following assumption, which is concerned with the stability of the projector P_h with respect to the norm $\|\cdot\|_1$, will mainly be needed for the proof of Lemma 3.13(ii) below and, consequently, also for the proof of Theorem 3.14.

Assumption 3.5. For the given family $(S_h)_{h \in (0,1]}$ of finite dimensional subspaces of $\dot{H}^1$ there exists a constant C with

$$
\|P_h x\|_1 \leq C \|x\|_1 \quad \text{for all } x \in \dot{H}^1, \; h \in (0, 1]. \tag{3.17}
$$

We conclude this section with two examples which satisfy Assumptions 3.3 and 3.5.

Example 3.6 (Standard finite element method). Assume that $H = L^2(\mathscr{D})$, where $\mathscr{D} \subset \mathbb{R}^d$, $d \in \{1, 2, 3\}$, is a bounded, convex domain (a polygon if $d = 2$ or a polyhedron if $d = 3$). As in Example 2.22 let the operator A be given by $Au = -\nabla \cdot (a(x)\nabla u) + c(x)u$ with $c(x) \geq 0$ and $a(x) \geq a_0 > 0$ for $x \in \mathscr{D}$ with Dirichlet boundary conditions.

By $(\mathscr{T}_h)_{h \in (0,1]}$ we denote a regular family of partitions of $\mathscr{D}$ into simplices, where h is the maximal meshsize. We define S_h to be the space of all continuous functions $y_h : \overline{\mathscr{D}} \to \mathbb{R}$, which are piecewise linear on $\mathscr{T}_h$ and zero on the boundary $\partial \mathscr{D}$. Then $S_h \subset \dot{H}^1$ and Assumption 3.3 holds by Thomée [66, Lemma 1.1] or Brenner and Scott [7, Theorem 5.4.8].

Further, if the family $(\mathscr{T}_h)_{h \in (0,1]}$ is quasi-uniform, then also Assumption 3.5 is satisfied. For a more detailed discussion of Assumption 3.5 in the context of the finite element method we refer to [9, 10, 16].

Example 3.7 (Spectral Galerkin method). In the same situation as in Example 3.6 we further assume that $\mathscr{D} = [0, 1] \subset \mathbb{R}$ and $-A$ is the Laplace operator with homogeneous Dirichlet boundary conditions. In this situation the orthonormal eigenfunctions and eigenvalues of the Laplace operator are explicitly known to be

$$\lambda_k = k^2\pi^2 \text{ and } e_k(y) = \sqrt{2}\sin(k\pi y) \text{ for all } k \in \mathbb{N}, \ k \geq 1, \ y \in \mathcal{D}.$$

For $N \in \mathbb{N}$ set $h := \lambda_{N+1}^{-\frac{1}{2}}$ and define

$$S_h := \text{span}\{e_k : k = 1, \ldots, N\}.$$

Note that $S_h \subset \dot{H}^r$ for every $r \in \mathbb{R}$. For $x \in \dot{H}^1$ we represent its Ritz projection $R_h x \in S_h$ in terms of the basis $(e_k)_{k=1}^N$. This yields $R_h x = \sum_{k=1}^N x_k^h e_k$, where the coefficients $(x_k^h)_{k=1}^N$ are given by

$$x_k^h = (R_h x, e_k) = \frac{1}{\lambda_k}(R_h x, Ae_k) = \frac{1}{\lambda_k}(R_h x, e_k)_1 = \frac{1}{\lambda_k}(x, e_k)_1 = (x, e_k).$$

Hence, in this example, the Ritz projector R_h is the restriction of the orthogonal L^2-projector P_h to $\dot{H}^1$. Moreover, we have

$$\left\| (\text{Id}_H - R_h)x \right\|^2 = \left\| (\text{Id}_H - P_h)x \right\|^2 = \sum_{k=N+1}^{\infty} (x, e_k)^2 = \sum_{k=N+1}^{\infty} \lambda_k^{-\rho}(x, A^{\frac{\rho}{2}}e_k)^2$$

$$\leq \lambda_{N+1}^{-\rho} \sum_{k=N+1}^{\infty} (A^{\frac{\rho}{2}}x, e_k)^2 = h^{2\rho}\|x\|_\rho^2 \text{ for all } x \in \dot{H}^\rho, \ \rho = 1, 2,$$

since $\lambda_k^{-1} \leq \lambda_{N+1}^{-1} = h^2$ for all $k \geq N + 1$. Therefore, Assumption 3.3 is satisfied for the spectral Galerkin method.

That Assumption 3.5 holds is easily seen by

$$\left\| P_h x \right\|_1^2 = \left\| A^{\frac{1}{2}} \sum_{k=1}^N (x, e_k)e_k \right\|^2 = \sum_{k=1}^N (A^{\frac{1}{2}}x, e_k)^2 \leq \|x\|_1^2 \quad \text{for all } x \in \dot{H}^1.$$

A detailed presentation of spectral Galerkin methods is found in [37].

3.3 Sharp Estimates for Deterministic Problems: The Semidiscrete Case

This section extends error estimates from [66] for the discretization of the deterministic linear homogeneous equation

$$\frac{d}{dt}u(t) + Au(t) = 0, \quad t > 0, \quad \text{with } u(0) = x, \tag{3.18}$$

to non-smooth initial data $x \in \dot{H}^{-1}$. We also present suitable integral versions of these estimates which are crucial for the derivation of the optimal error estimates.

The following two lemmas are concerned with the operator $F_h(t) := E_h(t) P_h - E(t)$, $t \geq 0$. Note that $F_h(t)x$ can be seen as the error at time $t \geq 0$ between the weak solution u to (3.18) and u_h which solves the spatially semidiscrete equation

$$\frac{\mathrm{d}}{\mathrm{d}t} u_h(t) + A_h u_h(t) = 0, \quad t > 0, \quad \text{with } u_h(0) = P_h x,$$

for $x \in \dot{H}^{-1}$.

Lemma 3.8. *Under Assumption 3.3 the following estimates hold for the error operator F_h.*

(i) Let $0 \leq \nu \leq \mu \leq 2$. Then there exists a constant C such that

$$\|F_h(t)x\| \leq C h^\mu t^{-\frac{\mu-\nu}{2}} \|x\|_\nu \text{ for all } x \in \dot{H}^\nu, \ t > 0, \ h \in (0, 1].$$

(ii) Let $0 \leq \rho \leq 1$. Then there exists a constant C such that

$$\|F_h(t)x\| \leq C t^{-\frac{\rho}{2}} \|x\|_{-\rho} \text{ for all } x \in \dot{H}^{-\rho}, \ t > 0, \ h \in (0, 1].$$

(iii) Let $0 \leq \rho \leq 1$. Then there exists a constant C such that

$$\|F_h(t)x\| \leq C h^{2-\rho} t^{-1} \|x\|_{-\rho} \text{ for all } x \in \dot{H}^{-\rho}, \ t > 0, \ h \in (0, 1].$$

Proof. The proof of estimate *(i)* can be found in [66, Theorem 3.5].

In order to prove *(ii)* we first note that the case $\rho = 0$ is true by *(i)*. Lemma B.9*(i)* yields

$$\|E(t)x\| = \|A^{\frac{1}{2}} E(t) A^{-\frac{1}{2}} x\| \leq C t^{-\frac{1}{2}} \|x\|_{-1}. \tag{3.19}$$

Together with (3.16) this proves

$$\|F_h(t)x\| \leq \|E_h(t) P_h x\| + \|E(t)x\| \leq C t^{-\frac{1}{2}} \|x\|_{-1}$$

for all $x \in \dot{H}^{-1}$. This settles the case $\rho = 1$. The intermediate cases follow by the interpolation technique which is demonstrated in the proof of [66, Theorem 3.5].

For *(iii)* the case $\rho = 0$ is again covered by *(i)*. Thus, it is enough to consider the case $\rho = 1$. First, by using (3.14), (3.12), and (3.8), we observe that

$$\|F_h(t)x\| = \|A_h E_h(t) A_h^{-1} P_h x - A E(t) A^{-1} x\|$$

$$\leq \|A_h E_h(t) P_h (R_h A^{-1} x - A^{-1} x)\| + \|(A_h E_h(t) P_h - A E(t)) A^{-1} x\|$$

$$\leq Ct^{-1}\|(R_h - \mathrm{Id}_H)A^{-1}x\| + \left\|\frac{\mathrm{d}F_h}{\mathrm{d}t}(t)A^{-1}x\right\|$$

$$\leq Ct^{-1}h\|A^{-1}x\|_1 + \left\|\frac{\mathrm{d}F_h}{\mathrm{d}t}(t)A^{-1}x\right\|.$$

Since $\|A^{-1}x\|_1 = \|x\|_{-1}$ the first term is already in the desired form. The last term is estimated by a slightly modified version of [66, Theorem 3.4], which gives

$$\left\|\frac{\mathrm{d}F_h}{\mathrm{d}t}(t)A^{-1}x\right\| \leq Cht^{-1}\|A^{-1}x\|_1.$$

This proves the case $\rho = 1$ and the intermediate cases follow by interpolation.   □

Lemma 3.9. *Let $0 \leq \rho \leq 1$. Under Assumption 3.3 the operator F_h satisfies the following estimates:*

(i) There exists a constant C such that

$$\left\|\int_0^t F_h(\sigma)x\,\mathrm{d}\sigma\right\| \leq Ch^{2-\rho}\|x\|_{-\rho} \text{ for all } x \in \dot{H}^{-\rho},\ t > 0,\ h \in (0,1].$$

(ii) There exists a constant C such that

$$\left(\int_0^t \|F_h(\sigma)x\|^2\,\mathrm{d}\sigma\right)^{\frac{1}{2}} \leq Ch^{1+\rho}\|x\|_\rho \text{ for all } x \in \dot{H}^\rho,\ t > 0,\ h \in (0,1].$$

Proof. As in the proof of the previous lemma it is enough to show the estimates for $\rho = 0$ and $\rho = 1$. Then the intermediate cases follow by interpolation.

The proof of *(i)* with $\rho = 0$ is contained in the proof of [66, Theorem 3.3], where the notation

$$\tilde{e}(t) = \int_0^t F_h(\sigma)x\,\mathrm{d}\sigma$$

is used.

Here we present a proof of *(i)* with $\rho = 1$. To this end we use (3.14) and find the estimate

$$\left\|\int_0^t F_h(\sigma)x\,\mathrm{d}\sigma\right\| = \left\|\int_0^t \left(A_h E_h(\sigma)A_h^{-1}P_h - AE(\sigma)A^{-1}\right)x\,\mathrm{d}\sigma\right\|$$

$$\leq \left\|\int_0^t A_h E_h(\sigma)P_h\left(R_h - \mathrm{Id}_H\right)A^{-1}x\,\mathrm{d}\sigma\right\|$$

$$+ \left\|\int_0^t \left(A_h E_h(\sigma)P_h - AE(\sigma)\right)A^{-1}x\,\mathrm{d}\sigma\right\|$$

$$= \left\| \int_0^t \frac{dE_h}{d\sigma}(\sigma) P_h (R_h - \mathrm{Id}_H) A^{-1} x \, d\sigma \right\|$$

$$+ \left\| \int_0^t \frac{dF_h}{d\sigma}(\sigma) A^{-1} x \, d\sigma \right\|.$$

By the fundamental theorem of calculus, $\|P_h y\| \le \|y\|$ for all $y \in H$, and Assumption 3.3 we have for the first term

$$\left\| \int_0^t \frac{dE_h}{d\sigma}(\sigma) P_h (R_h - \mathrm{Id}_H) A^{-1} x \, d\sigma \right\| = \left\| (E_h(t) - \mathrm{Id}_H) P_h (R_h - \mathrm{Id}_H) A^{-1} x \right\|$$

$$\le Ch \|A^{-1}x\|_1 = Ch \|x\|_{-1}.$$

For the second term we use Lemma 3.8(i) with $\mu = \nu = 1$. This yields

$$\left\| \int_0^t \frac{dF_h}{d\sigma}(\sigma) A^{-1} x \, d\sigma \right\| = \left\| (F_h(t) - F_h(0)) A^{-1} x \right\|$$

$$\le \left\| F_h(t) A^{-1} x \right\| + \left\| (\mathrm{Id}_H - P_h) A^{-1} x \right\| \le Ch \|x\|_{-1}.$$

In the last step we used the best approximation property of the orthogonal projector P_h, which, together with (3.8), gives

$$\left\| (P_h - \mathrm{Id}_H) y \right\| \le \left\| (R_h - \mathrm{Id}_H) y \right\| \le Ch \|y\|_1 \text{ for all } y \in \dot{H}^1.$$

It remains to prove (ii). From [66, (2.29)] we have the inequality

$$\int_0^t \left\| F_h(\sigma) x \right\|^2 d\sigma \le \int_0^t \left\| (R_h - \mathrm{Id}_H) E(\sigma) x \right\|^2 d\sigma.$$

In both cases, $\rho \in \{0, 1\}$, we have by (3.8)

$$\left\| (R_h - \mathrm{Id}_H) E(\sigma) x \right\| \le Ch^{1+\rho} \left\| E(\sigma) x \right\|_{1+\rho} = Ch^{1+\rho} \left\| A^{\frac{1}{2}} E(\sigma) A^{\frac{\rho}{2}} x \right\|.$$

Thus, if we apply Lemma B.9(iii) the proof is complete.   □

3.4 Spatially Semidiscrete Approximation of the Stochastic Equation

Our first main result of this chapter is concerned with the so called spatially semidiscrete approximation of (3.2), that is, we only discretize with respect to the Hilbert space H.

As in the previous section let $(S_h)_{h \in (0,1]}$ denote a family of finite dimensional subspaces of $\dot{H}^1$. We consider the uniquely determined stochastic process

$X_h\colon [0, T] \times \Omega \to S_h$ which solves the stochastic evolution equation

$$\mathrm{d}X_h(t) + [A_h X_h(t) + P_h f(t, X_h(t))]\,\mathrm{d}t = P_h g(t, X_h(t))\,\mathrm{d}W(t), \text{ for } 0 \le t \le T,$$
$$X_h(0) = P_h X_0,$$
$$(3.20)$$

where P_h denotes the generalized orthogonal projector onto S_h and $A_h\colon S_h \to S_h$ is the discrete version of the operator A from (3.11).

As for the continuous problem (3.1) there exists a unique mild solution X_h to (3.20) which satisfies

$$X_h(t) = E_h(t) P_h X_0 - \int_0^t E_h(t - \sigma) P_h f(\sigma, X_h(\sigma))\,\mathrm{d}\sigma$$

$$+ \int_0^t E_h(t - \sigma) P_h g(\sigma, X_h(\sigma))\,\mathrm{d}W(\sigma) \text{ for } 0 \le t \le T. \quad (3.21)$$

Before we formulate the convergence result let us recall the meaning of the two most crucial parameters. First, we have the parameter $r \in [0, 1)$ which controls the spatial regularity of the mild solution X as we have seen in Sect. 2.5. On the other hand, the parameter $h \in (0, 1]$ governs the granularity of the spatial approximation as in Examples 3.6 and 3.7.

According to the following theorem the semidiscrete approximation converges with order $1 + r$ to the mild solution X. Since this rate coincides with the spatial regularity of X it is called optimal (see [66, Chap. 1]).

Theorem 3.10. *Under the conditions of Sect. 2.3, Assumption 3.1 with $r \in [0, 1)$, $p \in [2, \infty)$, and Assumption 3.3 there exists a constant C, independent of $h \in (0, 1]$, such that*

$$\|X_h(t) - X(t)\|_{L^p(\Omega;H)} \le C h^{1+r}, \text{ for all } t \in (0, T],$$

where X_h and X denote the mild solutions to (3.20) and (3.1), respectively.

Proof. For $t \in (0, T]$ we have by (3.2) and (3.21)

$$\|X_h(t) - X(t)\|_{L^p(\Omega;H)} \le \|F_h(t) X_0\|_{L^p(\Omega;H)}$$

$$+ \left\| \int_0^t E_h(t - \sigma) P_h f(\sigma, X_h(\sigma))\,\mathrm{d}\sigma \right.$$

$$\left. - \int_0^t E(t - \sigma) f(\sigma, X(\sigma))\,\mathrm{d}\sigma \right\|_{L^p(\Omega;H)} \quad (3.22)$$

$$+ \left\| \int_0^t E_h(t - \sigma) P_h g(\sigma, X_h(\sigma))\,\mathrm{d}W(\sigma) \right.$$

$$\left. - \int_0^t E(t - \sigma) g(\sigma, X(\sigma))\,\mathrm{d}W(\sigma) \right\|_{L^p(\Omega;H)},$$

where $F_h(t) = E_h(t)P_h - E(t)$. The first term is estimated by Lemma 3.8(i) with $\mu = \nu = 1 + r$, which yields

$$\left\| F_h(t)X_0 \right\|_{L^p(\Omega;H)} \le Ch^{1+r} \left\| A^{\frac{1+r}{2}} X_0 \right\|_{L^p(\Omega;H)}. \tag{3.23}$$

The second term in (3.22) is dominated by three additional terms as follows

$$\left\| \int_0^t E_h(t-\sigma)P_h f(\sigma, X_h(\sigma)) \, d\sigma - \int_0^t E(t-\sigma) f(\sigma, X(\sigma)) \, d\sigma \right\|_{L^p(\Omega;H)}$$

$$\le \left\| \int_0^t E_h(t-\sigma)P_h \big(f(\sigma, X_h(\sigma)) - f(\sigma, X(\sigma)) \big) \, d\sigma \right\|_{L^p(\Omega;H)}$$

$$+ \left\| \int_0^t \big(E_h(t-\sigma)P_h - E(t-\sigma) \big) \big(f(\sigma, X(\sigma)) - f(t, X(t)) \big) \, d\sigma \right\|_{L^p(\Omega;H)}$$

$$+ \left\| \int_0^t \big(E_h(t-\sigma)P_h - E(t-\sigma) \big) f(t, X(t)) \, d\sigma \right\|_{L^p(\Omega;H)}$$

$$=: I_1 + I_2 + I_3.$$

We estimate each term separately. First note that, by (3.16), we have $\left\| E_h(t)P_h x \right\| \le Ct^{-\frac{1}{2}} \|x\|_{-1}$. Together with Lemma 2.26 this yields

$$I_1 \le \int_0^t \left\| E_h(t-\sigma)P_h \big(f(\sigma, X_h(\sigma)) - f(\sigma, X(\sigma)) \big) \right\|_{L^p(\Omega;H)} \, d\sigma$$

$$\le C \int_0^t (t-\sigma)^{-\frac{1}{2}} \left\| X_h(\sigma) - X(\sigma) \right\|_{L^p(\Omega;H)} \, d\sigma. \tag{3.24}$$

The term I_2 is estimated by applying Lemma 3.8(iii) with $\rho = 1 - r$ and Lemma 3.2. Then we get

$$I_2 \le \int_0^t \left\| F_h(t-\sigma) \big(f(\sigma, X(\sigma)) - f(t, X(t)) \big) \right\|_{L^p(\Omega;H)} \, d\sigma$$

$$\le Ch^{1+r} \int_0^t (t-\sigma)^{-1} \left\| f(\sigma, X(\sigma)) - f(t, X(t)) \right\|_{L^p(\Omega;\dot{H}^{-1+r})} \, d\sigma \tag{3.25}$$

$$\le Ch^{1+r} \int_0^t (t-\sigma)^{-1+\frac{1}{2}} \, d\sigma \left(1 + \sup_{\sigma \in [0,T]} \| X(\sigma) \|_{L^p(\Omega;\dot{H}^r)} \right).$$

The right hand side of this estimate is finite in view of (3.3).

Finally, the estimate for I_3 is a straightforward application of Lemma 3.9(i) with $\rho = 1 - r$. A further application of (3.7) gives

$$I_3 \leq Ch^{1+r} \| f(t, X(t)) \|_{L^p(\Omega; \dot{H}^{-1+r})} \leq Ch^{1+r} \Big(1 + \sup_{\sigma \in [0,T]} \| X(\sigma) \|_{L^p(\Omega; \dot{H}^r)} \Big).$$

$$(3.26)$$

A combination of the estimates (3.24)–(3.26) yields

$$\Big\| \int_0^t E_h(t - \sigma) P_h f(\sigma, X_h(\sigma)) \, \mathrm{d}\sigma - \int_0^t E(t - \sigma) f(\sigma, X(\sigma)) \, \mathrm{d}\sigma \Big\|_{L^p(\Omega; H)}$$

$$\leq Ch^{1+r} + C \int_0^t (t - \sigma)^{-\frac{1}{2}} \| X_h(\sigma) - X(\sigma) \|_{L^p(\Omega; H)} \, \mathrm{d}\sigma.$$

$$(3.27)$$

Next, we estimate the norm of the stochastic integral in (3.22). First, we apply Proposition 2.12 and get

$$\Big\| \int_0^t E_h(t - \sigma) P_h g(\sigma, X_h(\sigma)) \, \mathrm{d}W(\sigma) - \int_0^t E(t - \sigma) g(\sigma, X(\sigma)) \, \mathrm{d}W(\sigma) \Big\|_{L^p(\Omega; H)}$$

$$\leq C \Big(\mathbf{E} \Big[\Big(\int_0^t \| E_h(t - \sigma) P_h g(\sigma, X_h(\sigma)) - E(t - \sigma) g(\sigma, X(\sigma)) \|_{L_2^0}^2 \, \mathrm{d}\sigma \Big)^{\frac{p}{2}} \Big] \Big)^{\frac{1}{p}}.$$

The right hand side is a norm. Hence, the triangle inequality gives

$$\Big(\mathbf{E} \Big[\Big(\int_0^t \| E_h(t - \sigma) P_h g(\sigma, X_h(\sigma)) - E(t - \sigma) g(\sigma, X(\sigma)) \|_{L_2^0}^2 \, \mathrm{d}\sigma \Big)^{\frac{p}{2}} \Big] \Big)^{\frac{1}{p}}$$

$$\leq \Big\| \Big(\int_0^t \| E_h(t - \sigma) P_h \big(g(\sigma, X_h(\sigma)) - g(\sigma, X(\sigma)) \big) \|_{L_2^0}^2 \, \mathrm{d}\sigma \Big)^{\frac{1}{2}} \Big\|_{L^p(\Omega; \mathbb{R})}$$

$$+ \Big\| \Big(\int_0^t \| F_h(t - \sigma) \big(g(\sigma, X(\sigma)) - g(t, X(t)) \big) \|_{L_2^0}^2 \, \mathrm{d}\sigma \Big)^{\frac{1}{2}} \Big\|_{L^p(\Omega; \mathbb{R})}$$

$$+ \Big\| \Big(\int_0^t \| F_h(t - \sigma) g(t, X(t)) \|_{L_2^0}^2 \, \mathrm{d}\sigma \Big)^{\frac{1}{2}} \Big\|_{L^p(\Omega; \mathbb{R})}$$

$$=: I_4 + I_5 + I_6.$$

In a similar way as for I_1, we find an estimate for I_4 by using the stability of the operator $E_h(t) P_h$, that is, (3.12) with $\rho = 0$. Together with Lemma 2.26 we get

$$I_4 \leq C \Big\| \Big(\int_0^t \| g(\sigma, X_h(\sigma)) - g(\sigma, X(\sigma)) \|_{L_2^0}^2 \, \mathrm{d}\sigma \Big)^{\frac{1}{2}} \Big\|_{L^p(\Omega; \mathbb{R})}$$

$$\leq C \Big(\int_0^t \| g(\sigma, X_h(\sigma)) - g(\sigma, X(\sigma)) \|_{L^p(\Omega; L_2^0)}^2 \, \mathrm{d}\sigma \Big)^{\frac{1}{2}}$$

$$(3.28)$$

$$\leq C \Big(\int_0^t \| X_h(\sigma) - X(\sigma) \|_{L^p(\Omega; H)}^2 \, \mathrm{d}\sigma \Big)^{\frac{1}{2}}.$$

For the estimate of term I_5 we apply Lemma 3.8(i) with $\mu = 1 + r$ and $\nu = 0$, which gives $\|F_h(t)\| \le Ch^{1+r}t^{-\frac{1+r}{2}}$. The estimate is then completed by making use of Lemma 2.26 and the Hölder-continuity (3.4) with $s = 0$. Altogether, we derive

$$
\begin{aligned}
I_5 &\le Ch^{1+r} \left\| \left(\int_0^t (t - \sigma)^{-1-r} \|g(\sigma, X(\sigma)) - g(t, X(t))\|^2 \, d\sigma \right)^{\frac{1}{2}} \right\|_{L^p(\Omega;\mathbb{R})} \\
&\le Ch^{1+r} \left(\int_0^t (t - \sigma)^{-1-r} \|g(\sigma, X(\sigma)) - g(t, X(t))\|_{L^p(\Omega;H)}^2 \, d\sigma \right)^{\frac{1}{2}} \qquad (3.29) \\
&\le Ch^{1+r} \left(\int_0^t (t - \sigma)^{-1-r+1} \, d\sigma \right)^{\frac{1}{2}} \le Ch^{1+r},
\end{aligned}
$$

since $-r > -1$ and we therefore have

$$
\int_0^t (t - \sigma)^{-r} \, d\sigma = t^{1-r} \le T^{1-r}.
$$

Note that in the last step the generic constant C depends on $r \in [0, 1)$ and blows up as $r \to 1$.

Finally, for I_6, let $(\varphi_m)_{m\ge 1}$ denote an arbitrary orthonormal basis of the Hilbert space U_0. Then, by using Lemma 3.9(ii) with $\rho = r$, Assumption 2.20 and (3.3), we get

$$
\begin{aligned}
I_6 &= \left\| \left(\sum_{m=1}^\infty \int_0^t \|F_h(t - \sigma)g(t, X(t))\varphi_m\|^2 \, d\sigma \right)^{\frac{1}{2}} \right\|_{L^p(\Omega;\mathbb{R})} \\
&\le Ch^{1+r} \left\| \left(\sum_{m=1}^\infty \|A^{\frac{r}{2}}g(t, X(t))\varphi_m\|^2 \right)^{\frac{1}{2}} \right\|_{L^p(\Omega;\mathbb{R})} \qquad (3.30) \\
&= Ch^{1+r} \left\| \|g(t, X(t))\|_{L_{2,r}^0} \right\|_{L^p(\Omega;\mathbb{R})} \\
&\le Ch^{1+r} \left(1 + \sup_{\sigma \in [0,T]} \|X(\sigma)\|_{L^p(\Omega;\dot{H}^r)} \right) \le Ch^{1+r}.
\end{aligned}
$$

In total, we have by (3.28)–(3.30) that

$$
\left\| \int_0^t E_h(t - \sigma)P_h g(\sigma, X_h(\sigma)) \, dW(\sigma) - \int_0^t E(t - \sigma)g(\sigma, X(\sigma)) \, dW(\sigma) \right\|_{L^p(\Omega;H)}
$$

$$
\le Ch^{1+r} + C \left(\int_0^t \|X_h(\sigma) - X(\sigma)\|_{L^p(\Omega;H)}^2 \, d\sigma \right)^{\frac{1}{2}}.
$$

$$
(3.31)
$$

Coming back to (3.22), by (3.23), (3.27), and (3.31) we conclude that

$$\|X_h(t) - X(t)\|^2_{L^p(\Omega;H)} \leq Ch^{2(1+r)} + C \int_0^t \|X_h(\sigma) - X(\sigma)\|^2_{L^p(\Omega;H)}\, d\sigma$$

$$+ C\left(\int_0^t (t-\sigma)^{-\frac{1}{2}}\|X_h(\sigma) - X(\sigma)\|_{L^p(\Omega;H)}\, d\sigma\right)^2.$$

Finally, we note that

$$\int_0^t \|X_h(\sigma) - X(\sigma)\|^2_{L^p(\Omega;H)}\, d\sigma$$

$$= \int_0^t (t-\sigma)^{\frac{1}{2}}(t-\sigma)^{-\frac{1}{2}}\|X_h(\sigma) - X(\sigma)\|^2_{L^p(\Omega;H)}\, d\sigma$$

$$\leq T^{\frac{1}{2}}\int_0^t (t-\sigma)^{-\frac{1}{2}}\|X_h(\sigma) - X(\sigma)\|^2_{L^p(\Omega;H)}\, d\sigma,$$

and by Hölder's inequality

$$\int_0^t (t-\sigma)^{-\frac{1}{2}}\|X_h(\sigma) - X(\sigma)\|_{L^p(\Omega;H)}\, d\sigma$$

$$= \int_0^t (t-\sigma)^{-\frac{1}{4}}(t-\sigma)^{-\frac{1}{4}}\|X_h(\sigma) - X(\sigma)\|_{L^p(\Omega;H)}\, d\sigma$$

$$\leq \left(2T^{\frac{1}{2}}\right)^{\frac{1}{2}}\left(\int_0^t (t-\sigma)^{-\frac{1}{2}}\|X_h(\sigma) - X(\sigma)\|^2_{L^p(\Omega;H)}\, d\sigma\right)^{\frac{1}{2}}.$$

Hence, for $\varphi(t) = \|X_h(t) - X(t)\|^2_{L^p(\Omega;H)}$ we have shown that

$$\varphi(t) \leq Ch^{2(1+r)} + C\int_0^t (t-\sigma)^{-\frac{1}{2}}\varphi(\sigma)\, d\sigma$$

and the Gronwall Lemma A.2 completes the proof.   □

Remark 3.11. It remains to answer the question why we need to invoke the stronger
Assumption 3.1 in place of the linear growth condition on f from Assumption 2.19.

   The reason for this can, for example, be seen in the estimate (3.25), where we
applied Lemma 3.8(*iii*). Assumptions 2.14 and 2.19 would be sufficient to obtain a
similar estimate if instead of Lemma 3.8(*iii*) we have an estimate of the form

$$\|F_h(t)x\| \leq Ch^{1+r}t^{-1-\frac{r}{2}}\|x\|_{-1}, \quad r \in [0,1], \ x \in \dot{H}^{-1}.$$

It appears that such an estimate is not available in our general situation. However,
it is true, for example, for the spectral Galerkin method or for finite element spaces
consisting of piecewise polynomial functions of degree 2 or higher. It remains

an open question if Assumption 3.1 is also superfluous for piecewise linear finite elements.

3.5 Sharp Estimates for Deterministic Problems: The Fully Discrete Case

In this section we extend the results of Sect. 3.3 to a fully discrete approximation of the homogeneous equation

$$\frac{d}{dt}u(t) + Au(t) = 0, \quad t > 0, \quad \text{with } u(0) = x. \tag{3.32}$$

Our method of choice is a combination of the spatial Galerkin discretization together with the well-known implicit Euler scheme.

As in Sect. 3.3 let a family of subspaces $(S_h)_{h \in (0,1]} \subset \dot{H}^1$ be given. The fully discrete approximation scheme is defined by the recursion

$$U_h^j + kA_h U_h^j = U_h^{j-1}, \quad j = 1, 2, \ldots \quad \text{with } U_h^0 = P_h x. \tag{3.33}$$

Here $k \in (0, 1]$ denotes a fixed time step and $U_h^j \in S_h$ denotes the approximation of $u(t_j)$ at time $t_j = jk$. A closed form representation of (3.33) is given by

$$U_h^j = (I + kA_h)^{-j} P_h x, \quad j = 0, 1, 2, \ldots . \tag{3.34}$$

In order to make the results from [66, Chap. 7] accessible and to indicate generalizations to other one-step methods we introduce the rational function

$$R(z) = \frac{1}{1 + z} \quad \text{for } z \in \mathbb{R}, z \neq 1.$$

By $R(kA_h)$ we denote the linear operator which is defined by

$$R(kA_h)x = \sum_{m=1}^{N_h} R(k\lambda_{h,m})(x, \varphi_{h,m})\varphi_{h,m}, \tag{3.35}$$

where $(\lambda_{h,m})_{m=1}^{N_h}$ are the positive eigenvalues of $A_h \colon S_h \to S_h$ with corresponding orthonormal eigenvectors $(\varphi_{h,m})_{m=1}^{N_h} \subset S_h$ and $\dim(S_h) = N_h$. With this notation (3.33) is equivalently written as

$$U_h^j = R(kA_h)^j P_h x, \quad j = 0, 1, 2, \ldots . \tag{3.36}$$

The characteristic function R of the implicit Euler scheme enjoys the following properties with $q = 1$:

$$R(z) = e^{-z} + \mathcal{O}(z^{q+1}) \quad \text{for } z \to 0,$$
$$|R(z)| < 1 \quad \text{for all } z > 0, \quad \text{and } \lim_{z \to \infty} R(z) = 0. \tag{3.37}$$

In the nomenclature of [66, Chap. 7] the rational function $R(z)$ is an approximation of e^{-z} with accuracy $q = 1$ and is said to be of type IV. A one-step scheme, whose characteristic rational function possesses the properties (3.37), is unconditionally stable and satisfies, for $\rho \in [0, 1]$, the discrete smoothing property

$$\|A_h^\rho R(kA_h)^j x_h\| \le Ct_j^{-\rho}\|x_h\| \quad \text{for all } j = 1, 2, \ldots \text{ and } x_h \in S_h, \tag{3.38}$$

where the constant $C = C(\rho)$ is independent of k, h and j. For a proof of (3.38) we refer to [66, Lemma 7.3].

Further, as in the proof of [66, Theorem 7.1] it follows from (3.37) that there exists a constant C with

$$|R(z) - e^{-z}| \le Cz^{q+1} \text{ for all } z \in [0, 1] \tag{3.39}$$

and there exists a constant $c \in (0, 1)$ with

$$|R(z)| \le e^{-cz} \text{ for all } z \in [0, 1]. \tag{3.40}$$

The rest of this subsection deals with estimates of the error between the discrete approximation U_h^j and the solution $u(t_j)$. For the error analysis in Sect. 3.6 it will be convenient to introduce an error operator

$$F_{kh}(t) := E_{kh}(t)P_h - E(t),$$

which is defined for all $t \ge 0$, where

$$E_{kh}(t) := R(kA_h)^j, \quad \text{if } t \in [t_{j-1}, t_j) \text{ for } j = 1, 2, \ldots . \tag{3.41}$$

The mapping $t \mapsto E_{kh}(t)$, and hence $t \mapsto F_{kh}(t)$, is right continuous with left limits. A simple consequence of (3.38) and (3.15) is the inequality

$$\left\|E_{kh}(t)P_h x\right\| = \left\|A_h^{\frac{1}{2}} R(kA_h)^j A_h^{-\frac{1}{2}} P_h x\right\| \le Ct_j^{-\frac{1}{2}}\|x\|_{-1} \le Ct^{-\frac{1}{2}}\|x\|_{-1}, \tag{3.42}$$

which holds for all $x \in \dot{H}^{-1}$, $h, k \in (0, 1]$ and $t > 0$ with $t \in [t_{j-1}, t_j)$, $j = 1, 2, \ldots$.

The following lemma is the time discrete analogue of Lemma 3.8.

Lemma 3.12. *Under Assumption 3.3 the following estimates hold for the error operator* F_{kh}.

(i) Let $0 \leq \nu \leq \mu \leq 2$. *Then there exists a constant* C *such that*

$$\left\| F_{kh}(t)x \right\| \leq C\left(h^{\mu} + k^{\frac{\mu}{2}}\right)t^{-\frac{\mu-\nu}{2}}\left\| x \right\|_{\nu} \text{ for all } x \in \dot{H}^{\nu}, \ t > 0, \ h,k \in (0,1].$$

(ii) Let $0 \leq \rho \leq 1$. *Then there exists a constant* C *such that*

$$\left\| F_{kh}(t)x \right\| \leq Ct^{-\frac{\rho}{2}}\left\| x \right\|_{-\rho} \text{ for all } x \in \dot{H}^{-\rho}, \ t > 0, \ h,k \in (0,1].$$

(iii) Let $0 \leq \rho \leq 1$. *Then there exists a constant* C *such that*

$$\left\| F_{kh}(t)x \right\| \leq C\left(h^{2-\rho} + k^{\frac{2-\rho}{2}}\right)t^{-1}\left\| x \right\|_{-\rho} \text{ for all } x \in \dot{H}^{-\rho}, \ t > 0, \ h,k \in (0,1].$$

Proof. (i) Let $t > 0$ be such that $t_{j-1} \leq t < t_j$ and $x \in \dot{H}^{\nu}$. Then we get

$$\left\| F_{kh}(t)x \right\| \leq \left\| \left(R(kA_h)^j P_h - E(t_j)\right)x \right\| + \left\| \left(E(t_j) - E(t)\right)x \right\|.$$

For the second summand we have by Lemma B.9(i) and (ii)

$$\left\| \left(E(t_j) - E(t)\right)x \right\| = \left\| A^{-\frac{\mu}{2}}\left(E(t_j - t) - \mathrm{Id}_H\right)A^{\frac{\mu-\nu}{2}}E(t)A^{\frac{\nu}{2}}x \right\|$$
$$\leq C(t_j - t)^{\frac{\mu}{2}}t^{-\frac{\mu-\nu}{2}}\left\| A^{\frac{\nu}{2}}x \right\| \leq Ck^{\frac{\mu}{2}}t^{-\frac{\mu-\nu}{2}}\left\| x \right\|_{\nu}.$$

Further, the first summand is the error between the exact solution u of (3.18) and the fully discrete scheme (3.36) at time t_j. For the case $\mu = \nu = 2$, [66, Theorem 7.8] gives the estimate

$$\left\| \left(R(kA_h)^j P_h - E(t_j)\right)x \right\| \leq C\left(h^2 + k\right)\left\| x \right\|_2.$$

By the stability of the numerical scheme, that is (3.38) with $\rho = 0$, we also have the case $\mu = \nu = 0$. Hence,

$$\left\| F_{kh}(t_j)x \right\| \leq C\left\| x \right\|, \tag{3.43}$$

and, as above, the constant C is independent of $h,k \in (0,1]$, $t_j > 0$, and x. The same interpolation technique, which is used in the proof of [66, Theorem 7.8], gives us the intermediate cases for $\mu = \nu$ and $\mu \in [0,2]$, that is

$$\left\| \left(R(kA_h)^j P_h - E(t_j)\right)x \right\| \leq C\left(h^{\mu} + k^{\frac{\mu}{2}}\right)\left\| x \right\|_{\mu}. \tag{3.44}$$

On the other hand, [66, Theorem 7.7] proves the case $\nu = 0$ and $\mu = 2$. Hence, we have

$$\left\|\left(R(kA_h)^j P_h - E(t_j)\right)x\right\| \le C\left(h^2 + k\right)t_j^{-1}\|x\|,$$

where the constant C is again independent of $h, k \in (0, 1]$, $t_j > 0$, and $x \in H$. An interpolation between this estimate and (3.43) shows

$$\left\|\left(R(kA_h)^j P_h - E(t_j)\right)x\right\| \le C\left(h^\mu + k^{\frac{\mu}{2}}\right)t_j^{-\frac{\mu}{2}}\|x\| \tag{3.45}$$

for all $\mu \in [0, 2]$. For fixed $\mu \in [0, 2]$ the proof of *(i)* is completed by an additional interpolation with respect to $\nu \in [0, \mu]$ between (3.44) and (3.45) and the fact that $t_j^{-\frac{\mu}{2}} \le t^{-\frac{\mu}{2}}$.

The proof of *(ii)* works analogously. The case $\rho = 0$ is true by (3.43) and the case $\rho = 1$ follows by (3.19) and (3.42), since

$$\left\|F_{kh}(t)x\right\| \le \left\|E_{kh}(t)P_h x\right\| + \left\|E(t)x\right\| \le Ct^{-\frac{1}{2}}\|x\|_{-1}.$$

The intermediate cases follow by interpolation.

For *(iii)* the case $\rho = 0$ is already included in *(i)* with $\mu = 2$ and $\nu = 0$. Thus, it remains to show the case $\rho = 1$. For $t > 0$ with $t_{j-1} \le t < t_j$ we have

$$\left\|F_{kh}(t)x\right\| \le \left\|\left(R(kA_h)^j - E_h(t_j)\right)P_h x\right\| + \left\|\left(E_h(t_j)P_h - E(t_j)\right)x\right\|$$
$$+ \left\|\left(E(t_j) - E(t)\right)x\right\| =: T_1 + T_2 + T_3.$$

As in (3.35) we denote by $(\lambda_{h,m})_{m=1}^{N_h}$ the positive eigenvalues of A_h with corresponding orthonormal eigenvectors $(\varphi_{h,m})_{m=1}^{N_h} \subset S_h$. For T_1 we use the expansion of $P_h x$ in terms of $(\varphi_{h,m})_{m=1}^{N_h}$. This yields

$$T_1^2 = \left\|\sum_{m=1}^{N_h} \lambda_{h,m}^{\frac{1}{2}}\left(R(k\lambda_{h,m})^j - e^{-\lambda_{h,m}t_j}\right)\left(P_h x, \lambda_{h,m}^{-\frac{1}{2}}\varphi_{h,m}\right)\varphi_{h,m}\right\|^2$$

$$= \sum_{m=1}^{N_h} \lambda_{h,m}\left|R(k\lambda_{h,m})^j - e^{-k\lambda_{h,m}j}\right|^2\left(A_h^{-\frac{1}{2}}P_h x, \varphi_{h,m}\right)^2.$$

First, we consider all summands with $k\lambda_{h,m} \le 1$. As in the proof of [66, Theorem 7.1], by applying (3.39) with $q = 1$ and (3.40), we get

$$\left|R(k\lambda_{h,m})^j - e^{-k\lambda_{h,m}j}\right| = \left|\left(R(k\lambda_{h,m}) - e^{-k\lambda_{h,m}}\right)\sum_{i=0}^{j-1} R(k\lambda_{h,m})^{j-1-i}e^{-k\lambda_{h,m}i}\right|$$

$$\le Cj\left(k\lambda_{h,m}\right)^2 e^{-c(j-1)k\lambda_{h,m}}.$$

$$\tag{3.46}$$

Therefore, since $t_j = jk$ and $k\lambda_{h,m} \leq 1$ it holds true that

$$\lambda_{h,m}\left|R(k\lambda_{h,m})^j - \mathrm{e}^{-k\lambda_{h,m}j}\right|^2 \leq C(jk)^{-2}k^2\lambda_{h,m}\left(jk\lambda_{h,m}\right)^4\mathrm{e}^{-2cjk\lambda_{h,m}}\mathrm{e}^{2ck\lambda_{h,m}}$$
$$\leq Ct_j^{-2}k,$$

where we also used that $\sup_{z\geq 0} z^4\mathrm{e}^{-2cz} < \infty$.

For all summands with $k\lambda_{h,m} > 1$ we get the estimate

$$\lambda_{h,m}\left|R(k\lambda_{h,m})^j - \mathrm{e}^{-k\lambda_{h,m}j}\right|^2 < 2k^{-1}\left(k\lambda_{h,m}\right)^2\left(\left|R(k\lambda_{h,m})^j\right|^2 + \left|\mathrm{e}^{-k\lambda_{h,m}j}\right|^2\right).$$

As it is shown in the proof of [66, Lemma 7.3], we have

$$|R(z)| \leq \frac{1}{1+cz}, \quad \text{for all } z \geq 1, \text{ with } c > 0. \tag{3.47}$$

In fact, for the implicit Euler scheme this is immediately true with $c = 1$, but it also holds for all rational functions $R(z)$, which satisfy (3.37).

Together with $k\lambda_{h,m} > 1$ this yields

$$\left|k\lambda_{h,m}R(k\lambda_{h,m})^j\right|^2 \leq \left(\frac{k\lambda_{h,m}}{1+ck\lambda_{h,m}}\right)^2\left(1+ck\lambda_{h,m}\right)^{-2(j-1)}$$
$$\leq \frac{1}{c^2}(1+c)^{-2(j-1)} = \frac{1}{c^2}\mathrm{e}^{-2(j-1)\log(1+c)} \leq Cj^{-2}.$$

As above we also have

$$\left|k\lambda_{h,m}\mathrm{e}^{-k\lambda_{h,m}j}\right|^2 \leq Cj^{-2}.$$

Therefore, also in the case $k\lambda_{h,m} > 1$, it holds that

$$\lambda_{h,m}\left|R(k\lambda_{h,m})^j - \mathrm{e}^{-k\lambda_{h,m}j}\right|^2 \leq Ct_j^{-2}k.$$

Together with Parseval's identity and (3.15) we arrive at

$$T_1^2 \leq Ct_j^{-2}k\sum_{m=1}^{\infty}\left(A_h^{-\frac{1}{2}}P_hx, \varphi_{h,m}\right)^2 = Ct_j^{-2}k\left\|A_h^{-\frac{1}{2}}P_hx\right\|^2 \leq Ct^{-2}k\|x\|_{-1}^2.$$

The term T_2 is covered by Lemma 3.8(*iii*) which gives

$$T_2 = \left\|F_h(t_j)x\right\| \leq Cht_j^{-1}\|x\|_{-1} \leq Cht^{-1}\|x\|_{-1}.$$

Finally, for T_3 we apply Lemma B.9(*i*) with $\nu = 1$ and (*ii*) with $\nu = \frac{1}{2}$ and get

$$T_3 = \left\| AE(t)A^{-\frac{1}{2}}\left(E(t_j - t) - \mathrm{Id}_H \right) A^{-\frac{1}{2}} x \right\|$$
$$\leq Ct^{-1}(t_j - t)^{\frac{1}{2}} \|x\|_{-1} \leq Ct^{-1}k^{\frac{1}{2}} \|x\|_{-1}.$$

Combining the estimates for T_1, T_2 and T_3 proves *(iii)* with $\rho = 1$. As usual, the intermediate cases follow by interpolation.   □

We also have an analogue of Lemma 3.9. A time discrete version of Lemma 3.13*(ii)*, where the integral is replaced by a sum, is shown in [70].

Lemma 3.13. *Let* $0 \leq \rho \leq 1$. *Under Assumption 3.3 the operator* F_{kh} *satisfies the following estimates:*

(i) There exists a constant C such that

$$\left\| \int_0^t F_{kh}(\sigma)x \, d\sigma \right\| \leq C\left(h^{2-\rho} + k^{\frac{2-\rho}{2}} \right) \|x\|_{-\rho}$$

for all $x \in \dot{H}^{-\rho}$, $t > 0$, *and* $h, k \in (0, 1]$.
(ii) Under the additional Assumption 3.5 there exists a constant C such that

$$\left(\int_0^t \left\| F_{kh}(\sigma)x \right\|^2 d\sigma \right)^{\frac{1}{2}} \leq C\left(h^{1+\rho} + k^{\frac{1+\rho}{2}} \right) \|x\|_{\rho}$$

for all $x \in \dot{H}^{\rho}$, $t > 0$, *and* $h, k \in (0, 1]$.

Proof. The proof of *(i)* uses a similar technique as the proof of Lemma 3.12*(iii)*. First, without loss of generality, we can assume that $t = t_n$ for some $n \geq 0$. In fact, if $t_n < t < t_{n+1}$ then we have

$$\left\| \int_0^t F_{kh}(\sigma)x \, d\sigma \right\| \leq \left\| \int_0^{t_n} F_{kh}(\sigma)x \, d\sigma \right\| + \left\| \int_{t_n}^t F_{kh}(\sigma)x \, d\sigma \right\|.$$

For the second term we get by Lemma 3.12*(iii)*

$$\left\| \int_{t_n}^t F_{kh}(\sigma)x \, d\sigma \right\| \leq \left\| \int_{t_n}^t \left(F_{kh}(\sigma) - F_{kh}(t) \right)x \, d\sigma \right\| + \left\| \int_{t_n}^t F_{kh}(t)x \, d\sigma \right\|$$

$$= \left\| \int_{t_n}^t \left(E(\sigma) - E(t) \right)x \, d\sigma \right\| + (t - t_n)\left\| F_{kh}(t)x \right\|$$

$$\leq \left\| E(t_n)A^{\frac{\rho}{2}} \int_{t_n}^t E(\sigma - t_n)A^{-\frac{\rho}{2}} x \, d\sigma \right\| + (t - t_n)\left\| A^{\frac{\rho}{2}} E(t)A^{-\frac{\rho}{2}} x \right\|$$

$$+ C(t - t_n)\left(h^{2-\rho} + k^{\frac{2-\rho}{2}} \right)t^{-1} \|x\|_{-\rho}.$$

We continue by applying Lemma B.9*(iv)* and *(i)* with $\mu = \frac{\rho}{2}$ and the fact that $(t - t_n)t^{-1} \leq 1$ which yields

$$\left\| \int_{t_n}^{t} F_{kh}(\sigma)x \, d\sigma \right\| \leq C \left((t - t_n)^{1 - \frac{\rho}{2}} + (t - t_n)t^{-\frac{\rho}{2}} + h^{2-\rho} + k^{\frac{2-\rho}{2}} \right) \|x\|_{-\rho}$$

$$\leq C \left(h^{2-\rho} + k^{\frac{2-\rho}{2}} \right) \|x\|_{-\rho}.$$

Next, we have

$$\left\| \int_{0}^{t_n} F_{kh}(\sigma)x \, d\sigma \right\| \leq \left\| \int_{0}^{t_n} \left(E_{kh}(\sigma) - E_h(\sigma) \right) P_h x \, d\sigma \right\|$$

$$+ \left\| \int_{0}^{t_n} \left(E_h(\sigma) P_h - E(\sigma) \right)x \, d\sigma \right\|.$$

For the second term Lemma 3.9*(i)* yields the bound

$$\left\| \int_{0}^{t_n} \left(E_h(\sigma) P_h - E(\sigma) \right)x \, d\sigma \right\| \leq C h^{2-\rho} \|x\|_{-\rho}.$$

Thus, it is enough to show that

$$\left\| \int_{0}^{t_n} \left(E_{kh}(\sigma) - E_h(\sigma) \right) P_h x \, d\sigma \right\| \leq C k^{\frac{2-\rho}{2}} \|x\|_{-\rho},$$

where the constant $C = C(\rho)$ is independent of $h, k \in (0, 1], t > 0$, and $x \in \dot{H}^{-\rho}$. We plug in the definition of E_{kh} and obtain

$$\left\| \int_{0}^{t_n} \left(E_{kh}(\sigma) - E_h(\sigma) \right) P_h x \, d\sigma \right\| \leq \left\| \sum_{j=1}^{n} \int_{t_{j-1}}^{t_j} \left(R(kA_h)^j - E_h(t_j) \right) P_h x \, d\sigma \right\|$$

$$+ \left\| \sum_{j=1}^{n} \int_{t_{j-1}}^{t_j} \left(E_h(t_j) - E_h(\sigma) \right) P_h x \, d\sigma \right\|.$$

$$(3.48)$$

As in (3.35) let $(\lambda_{h,m})_{m=1}^{N_h}$ be the positive eigenvalues of A_h with corresponding orthonormal eigenvectors $(\varphi_{h,m})_{m=1}^{N_h} \subset S_h$. Then, Parseval's identity yields for the first summand

$$\left\| \sum_{j=1}^{n} \int_{t_{j-1}}^{t_j} \left(R(kA_h)^j - E_h(t_j) \right) P_h x \, d\sigma \right\|^2$$

$$= \sum_{m=1}^{N_h} \left| k \sum_{j=1}^{n} \left(R(k\lambda_{h,m})^j - e^{-\lambda_{h,m} t_j} \right) \right|^2 \left(P_h x, \varphi_{h,m} \right)^2$$

$$\leq \sum_{m=1}^{N_h} \left(k \sum_{j=1}^{n} \lambda_{h,m}^{\frac{\varrho}{2}} \left| R(k\lambda_{h,m})^j - e^{-\lambda_{h,m} t_j} \right| \right)^2 \left(A_h^{-\frac{\varrho}{2}} P_h x, \varphi_{h,m} \right)^2.$$

As in the proof of Lemma 3.12(iii) we first study all summands with $k\lambda_{h,m} \leq 1$. In this case (3.46) gives

$$k \sum_{j=1}^{n} \lambda_{h,m}^{\frac{\varrho}{2}} \left| R(k\lambda_{h,m})^j - e^{-\lambda_{h,m} t_j} \right| \leq Ck \sum_{j=1}^{n} \lambda_{h,m}^{\frac{\varrho}{2}} j \left(k\lambda_{h,m} \right)^2 e^{-c(j-1)k\lambda_{h,m}}$$

$$= C\lambda_{h,m}^{\frac{\varrho+4}{2}} e^{ck\lambda_{h,m}} k^2 \sum_{j=1}^{n} j k e^{-cjk\lambda_{h,m}}$$

$$\leq C\lambda_{h,m}^{\frac{\varrho+4}{2}} k \int_{0}^{\infty} (\sigma + k) e^{-c\lambda_{h,m}\sigma} \, d\sigma$$

$$\leq C\lambda_{h,m}^{\frac{\varrho+4}{2}} k \left(\frac{1}{(c\lambda_{h,m})^2} + \frac{k}{c\lambda_{h,m}} \right) \leq Ck^{\frac{2-\varrho}{2}}.$$

For all summands with $k\lambda_{h,m} > 1$ we have the estimates

$$k \sum_{j=1}^{n} \lambda_{h,m}^{\frac{\varrho}{2}} \left| R(k\lambda_{h,m})^j - e^{-\lambda_{h,m} t_j} \right|$$

$$< k^{\frac{2-\varrho}{2}} \sum_{j=1}^{n} k\lambda_{h,m} \left(\left| R(k\lambda_{h,m}) \right|^j + e^{-k\lambda_{h,m} j} \right)$$

$$\leq k^{\frac{2-\varrho}{2}} \left(\frac{k\lambda_{h,m}}{1 + ck\lambda_{h,m}} \sum_{j=1}^{n} (1+c)^{-(j-1)} + k\lambda_{h,m} e^{-k\lambda_{h,m}} \sum_{j=1}^{n} e^{-(j-1)} \right) \leq Ck^{\frac{2-\varrho}{2}},$$

where we used (3.47) and $e^{-k\lambda_{h,m}(j-1)} < e^{-(j-1)}$ for $k\lambda_{h,m} > 1$. Altogether, this proves

$$\left\| \sum_{j=1}^{n} \int_{t_{j-1}}^{t_j} \left(R(kA_h)^j - E_h(t_j) \right) P_h x \, d\sigma \right\|^2$$

$$\leq Ck^{2-\varrho} \sum_{j=1}^{N_h} \left(A_h^{-\frac{\varrho}{2}} P_h x, \varphi_{h,m} \right)^2 = Ck^{2-\varrho} \left\| A_h^{-\frac{\varrho}{2}} P_h x \right\|^2. \tag{3.49}$$

In order to complete the proof of *(i)* it remains to find an estimate for the second term in (3.48). By applying Parseval's identity in the same way as above we get

$$\left\| \sum_{j=1}^{n} \int_{t_{j-1}}^{t_j} \left(E_h(t_j) - E_h(\sigma) \right) P_h x \, d\sigma \right\|^2$$

$$= \sum_{m=1}^{N_h} \left| \sum_{j=1}^{n} \int_{t_{j-1}}^{t_j} \left(e^{-\lambda_{h,m} t_j} - e^{-\lambda_{h,m} \sigma} \right) d\sigma \right|^2 \left(P_h x, \varphi_{h,m} \right)^2$$

$$= \sum_{m=1}^{N_h} \left| \lambda_{h,m}^{\frac{\rho}{2}} \sum_{j=1}^{n} e^{-\lambda_{h,m} t_{j-1}} \int_{0}^{k} \left(e^{-k\lambda_{h,m}} - e^{-\lambda_{h,m} \sigma} \right) d\sigma \right|^2 \left(A_h^{-\frac{\rho}{2}} P_h x, \varphi_{h,m} \right)^2.$$

Since it holds that

$$\int_{0}^{k} \left(e^{-k\lambda_{h,m}} - e^{-\lambda_{h,m} \sigma} \right) d\sigma = k e^{-k\lambda_{h,m}} - \frac{1}{\lambda_{h,m}} \left(1 - e^{-k\lambda_{h,m}} \right),$$

we have

$$\left| \lambda_{h,m}^{\frac{\rho}{2}} \sum_{j=1}^{n} e^{-\lambda_{h,m} t_{j-1}} \int_{0}^{k} \left(e^{-k\lambda_{h,m}} - e^{-\lambda_{h,m} \sigma} \right) d\sigma \right|^2$$

$$= \left| \lambda_{h,m}^{\frac{\rho}{2}} \left(k e^{-k\lambda_{h,m}} - \frac{1}{\lambda_{h,m}} \left(1 - e^{-k\lambda_{h,m}} \right) \right) \sum_{j=1}^{n} e^{-k\lambda_{h,m}(j-1)} \right|^2$$

$$= \lambda_{h,m}^{\rho-2} \left| k \lambda_{h,m} e^{-k\lambda_{h,m}} - \left(1 - e^{-k\lambda_{h,m}} \right) \right|^2 \frac{(1 - e^{-\lambda_{h,m} t_n})^2}{\left(1 - e^{-k\lambda_{h,m}} \right)^2}.$$

We estimate $(1 - e^{-\lambda_{h,m} t_n})^2 \leq 1$. Further, if $k\lambda_{h,m} \leq 1$ then it is true that

$$\left| k\lambda_{h,m} e^{-k\lambda_{h,m}} - \left(1 - e^{-k\lambda_{h,m}} \right) \right|^2 = e^{-2k\lambda_{h,m}} \left| e^{k\lambda_{h,m}} - 1 - k\lambda_{h,m} \right|^2$$

$$\leq Ck^4 \lambda_{h,m}^4.$$

Thus, in this case we derive the estimate

$$\left| \lambda_{h,m}^{\frac{\rho}{2}} \sum_{j=1}^{n} e^{-\lambda_{h,m} t_{j-1}} \int_{0}^{k} \left(e^{-k\lambda_{h,m}} - e^{-\lambda_{h,m} \sigma} \right) d\sigma \right|^2$$

$$\leq Ck^{2-\rho} (k\lambda_{h,m})^{\rho} \frac{\lambda_{h,m}^2 k^2}{\left(1 - e^{-k\lambda_{h,m}} \right)^2} \leq Ck^{2-\rho},$$

where we have used that the function $x \mapsto x(1-e^{-x})^{-1}$ is bounded for all $x \in (0, 1]$. On the other hand, if $k\lambda_{h,m} > 1$ then we have

$$\left| \lambda_{h,m}^{\frac{\rho}{2}} \sum_{j=1}^{n} e^{-\lambda_{h,m} t_{j-1}} \int_{0}^{k} \left(e^{-k\lambda_{h,m}} - e^{-\lambda_{h,m}\sigma} \right) d\sigma \right|^{2}$$

$$\leq 2\lambda_{h,m}^{\rho-2} \left(\left| k\lambda_{h,m} e^{-k\lambda_{h,m}} \right|^{2} + \left| 1 - e^{-k\lambda_{h,m}} \right|^{2} \right) \left(1 - e^{-k\lambda_{h,m}} \right)^{-2} \leq Ck^{2-\rho},$$

since $\lambda_{h,m}^{\rho-2} < k^{2-\rho}$, $\sup_{x \geq 0} xe^{-x} < \infty$ and $(1 - e^{-k\lambda_{h,m}})^{-2} \leq (1 - e^{-1})^{-2}$. Altogether, this yields

$$\left\| \sum_{j=1}^{n} \int_{t_{j-1}}^{t_{j}} \left(E_{h}(t_{j}) - E_{h}(\sigma) \right) P_{h} x \, d\sigma \right\|^{2} \leq Ck^{2-\rho} \sum_{m=1}^{N_{h}} \left(A_{h}^{-\frac{\rho}{2}} P_{h} x, \varphi_{h,m} \right)^{2}$$

$$\leq Ck^{2-\rho} \left\| A_{h}^{-\frac{\rho}{2}} P_{h} x \right\|^{2},$$

which in combination with (3.49) and (3.15) completes the proof of *(i)* for $\rho \in \{0, 1\}$. The intermediate cases follow again by interpolation.

As above we begin the proof of *(ii)* with the remark that without loss of generality we can assume that $t = t_{n}$ for some $n \geq 0$. In a similar way as in the proof of *(i)* we have

$$\left(\int_{t_{n}}^{t} \left\| F_{kh}(\sigma)x \right\|^{2} d\sigma \right)^{\frac{1}{2}} \leq \left(\int_{t_{n}}^{t} \left\| (E(\sigma) - E(t))x \right\|^{2} d\sigma \right)^{\frac{1}{2}} + (t - t_{n})^{\frac{1}{2}} \left\| F_{kh}(t)x \right\|.$$

For the second term Lemma 3.12*(i)* with $\mu = 1 + \rho$ and $\nu = \rho$ together with $(t - t_{n})t^{-1} \leq 1$ gives the desired estimate. The first summand is estimated by Lemma B.9*(ii)* which gives

$$\left(\int_{t_{n}}^{t} \left\| (E(\sigma) - E(t))x \right\|^{2} d\sigma \right)^{\frac{1}{2}} = \left(\int_{t_{n}}^{t} \left\| E(\sigma)A^{-\frac{\rho}{2}} \left(\mathrm{Id}_{H} - E(t - \sigma) \right) A^{\frac{\rho}{2}} x \right\|^{2} d\sigma \right)^{\frac{1}{2}}$$

$$\leq \left(\int_{t_{n}}^{t} (t - \sigma)^{\rho} d\sigma \right)^{\frac{1}{2}} \| x \|_{\rho} \leq Ck^{\frac{1+\rho}{2}} \| x \|_{\rho}.$$

Further, we have

$$\left(\int_{0}^{t_{n}} \left\| F_{kh}(\sigma)x \right\|^{2} d\sigma \right)^{\frac{1}{2}} \leq \left(\int_{0}^{t_{n}} \left\| (E_{kh}(\sigma) - E_{h}(\sigma)) P_{h} x \right\|^{2} d\sigma \right)^{\frac{1}{2}}$$

$$+ \left(\int_{0}^{t_{n}} \left\| (E_{h}(\sigma) P_{h} - E(\sigma))x \right\|^{2} d\sigma \right)^{\frac{1}{2}}$$

and Lemma 3.9*(ii)* yields an estimate for the second summand of the form

$$\left(\int_{0}^{t_{n}} \left\| (E_{h}(\sigma) P_{h} - E(\sigma))x \right\|^{2} d\sigma \right)^{\frac{1}{2}} \leq Ch^{1+\rho} \| x \|_{\rho}.$$

Thus it remains to show

$$\left(\int_0^{t_n} \left\| (E_{kh}(\sigma) - E_h(\sigma)) P_h x \right\|^2 d\sigma \right)^{\frac{1}{2}} \le C k^{\frac{1+\rho}{2}} \|x\|_\rho.$$

As above, we prove this estimate for $\rho \in \{0, 1\}$. The intermediate cases follow again by interpolation. By the definition of E_{kh} we obtain the analogue of (3.48), namely

$$
\left(\int_0^{t_n} \left\| (E_{kh}(\sigma) - E_h(\sigma)) P_h x \right\|^2 d\sigma \right)^{\frac{1}{2}}
$$

$$
\le \left(\sum_{j=1}^n \int_{t_{j-1}}^{t_j} \left\| (R(kA_h)^j - E_h(t_j)) P_h x \right\|^2 d\sigma \right)^{\frac{1}{2}}
\tag{3.50}
$$

$$
+ \left(\sum_{j=1}^n \int_{t_{j-1}}^{t_j} \left\| (E_h(t_j) - E_h(\sigma)) P_h x \right\|^2 d\sigma \right)^{\frac{1}{2}}.
$$

For the square of the first summand we again apply Parseval's identity with respect to the orthonormal eigenbasis $(\varphi_{h,m})_{m=1}^{N_h} \subset S_h$ of A_h with corresponding eigenvalues $(\lambda_{h,m})_{m=1}^{N_h}$ and get

$$
\sum_{j=1}^n \int_{t_{j-1}}^{t_j} \left\| (R(kA_h)^j - E_h(t_j)) P_h x \right\|^2 d\sigma
$$

$$
= \sum_{j=1}^n k \sum_{m=1}^{N_h} \lambda_{h,m}^{-\rho} \left| R(k\lambda_{h,m})^j - e^{-k\lambda_{h,m}j} \right|^2 \left(A_{h,m}^{\frac{\rho}{2}} P_h x, \varphi_{h,m} \right)^2.
$$

For all summands with $k\lambda_{h,m} \le 1$ we apply (3.46). This yields

$$
\sum_{j=1}^n \int_{t_{j-1}}^{t_j} \left\| (R(kA_h)^j - E_h(t_j)) P_h x \right\|^2 d\sigma
$$

$$
\le C \sum_{m=1}^{N_h} k \lambda_{h,m}^{-\rho} \sum_{j=1}^n j^2 \left(k\lambda_{h,m} \right)^4 e^{-2c(j-1)k\lambda_{h,m}} \left(A_h^{\frac{\rho}{2}} P_h x, \varphi_{h,m} \right)^2
$$

$$
\le C \sum_{m=1}^{N_h} k^2 \lambda_{h,m}^{4-\rho} e^{2ck\lambda_{h,m}} \int_0^\infty (\sigma + k)^2 e^{-2c\lambda_{h,m}\sigma} d\sigma \left(A_h^{\frac{\rho}{2}} P_h x, \varphi_{h,m} \right)^2
\tag{3.51}
$$

$$
\le C \sum_{m=1}^{N_h} k^2 \lambda_{h,m}^{4-\rho} \left(\frac{2}{(2c\lambda_{h,m})^3} + \frac{2k}{(2c\lambda_{h,m})^2} + \frac{k^2}{2c\lambda_{h,m}} \right) \left(A_h^{\frac{\rho}{2}} P_h x, \varphi_{h,m} \right)^2
$$

$$
\le C k^{1+\rho} \left\| A_h^{\frac{\rho}{2}} P_h x \right\|^2.
$$

For the remaining summands with $k\lambda_{h,m} > 1$ we use (3.47) and get the estimate

$$\sum_{j=1}^{n} \left| R(k\lambda_{h,m})^j - e^{-k\lambda_{h,m}j} \right|^2 \le 2\sum_{j=1}^{n} \left((1+c)^{-2j} + e^{-2j} \right) \le C,$$

where the bound C is independent of n. Hence, also in this case we have

$$\sum_{j=1}^{n} \int_{t_{j-1}}^{t_j} \left\| \left(R(kA_h)^j - E_h(t_j) \right) P_h x \right\|^2 d\sigma$$

$$\le C \sum_{m=1}^{N_h} k\lambda_{h,m}^{-\rho} \left(A_h^{\frac{\rho}{2}} P_h x, \varphi_{h,m} \right)^2 < Ck^{1+\rho} \left\| A_h^{\frac{\rho}{2}} P_h x \right\|^2. \tag{3.52}$$

Next, we prove a similar result for the square of the second summand in (3.50). As in the proof of part *(i)* an application of Parseval's identity yields

$$\sum_{j=1}^{n} \int_{t_{j-1}}^{t_j} \left\| \left(E_h(t_j) - E_h(\sigma) \right) P_h x \right\|^2 d\sigma$$

$$= \sum_{j=1}^{n} \sum_{m=1}^{N_h} \lambda_{h,m}^{-\rho} \int_{t_{j-1}}^{t_j} \left| e^{-k\lambda_{h,m}j} - e^{-\lambda_{h,m}\sigma} \right|^2 d\sigma \left(A_h^{\frac{\rho}{2}} P_h x, \varphi_{h,m} \right)^2.$$

By using the fact

$$\int_{t_{j-1}}^{t_j} \left| e^{-k\lambda_{h,m}j} - e^{-\lambda_{h,m}\sigma} \right|^2 d\sigma \le e^{-2k\lambda_{h,m}(j-1)} k \left(1 - e^{-k\lambda_{h,m}} \right)^2$$

we obtain

$$\sum_{j=1}^{n} \int_{t_{j-1}}^{t_j} \left\| \left(E_h(t_j) - E_h(\sigma) \right) P_h x \right\|^2 d\sigma$$

$$\le \sum_{m=1}^{N_h} \lambda_{h,m}^{-\rho} k \left(1 - e^{-k\lambda_{h,m}} \right)^2 \left(A_h^{\frac{\rho}{2}} P_h x, \varphi_{h,m} \right)^2 \sum_{j=1}^{n} e^{-2k\lambda_{h,m}(j-1)}$$

$$\le \sum_{m=1}^{N_h} \lambda_{h,m}^{-\rho} k \left(1 - e^{-k\lambda_{h,m}} \right)^2 \left(A_h^{\frac{\rho}{2}} P_h x, \varphi_{h,m} \right)^2 \left(1 - e^{-2k\lambda_{h,m}} \right)^{-1}.$$

Since $1 - e^{-2k\lambda_{h,m}} = (1 + e^{-k\lambda_{h,m}})(1 - e^{-k\lambda_{h,m}})$ we conclude

$$\sum_{j=1}^{n} \int_{t_{j-1}}^{t_j} \left\| \left(E_h(t_j) - E_h(\sigma) \right) P_h x \right\|^2 \, d\sigma$$

$$\leq \sum_{m=1}^{N_h} \lambda_{h,m}^{-\rho} k \left(1 - e^{-k\lambda_{h,m}} \right) \left(A_h^{\frac{\ell}{2}} P_h x, \varphi_{h,m} \right)^2 \left(1 + e^{-k\lambda_{h,m}} \right)^{-1}$$

$$\leq \frac{1}{2} k \sum_{m=1}^{N_h} \lambda_{h,m}^{-\rho} \left(1 - e^{-k\lambda_{h,m}} \right) \left(A_h^{\frac{\ell}{2}} P_h x, \varphi_{h,m} \right)^2 .$$

If $\rho = 0$ this simplifies to $\frac{1}{2} k \| P_h x \|^2$ since $1 - e^{-k\lambda_{h,m}} \leq 1$. If $\rho = 1$ then we use $1 - e^{-k\lambda_{h,m}} \leq k\lambda_{h,m}$ and the right hand side is bounded by $\frac{1}{2} k^2 \| A_h^{\frac{1}{2}} P_h x \|^2$. Altogether, this proves

$$\sum_{j=1}^{n} \int_{t_{j-1}}^{t_j} \left\| \left(E_h(t_j) - E_h(\sigma) \right) P_h x \right\|^2 \, d\sigma \leq C k^{1+\rho} \| A_h^{\frac{1}{2}} P_h x \|^2 . \tag{3.53}$$

Finally, a combination of (3.50) with (3.51), (3.52) and (3.53) gives

$$\left(\int_{0}^{t_n} \left\| \left(E_{kh}(\sigma) - E_h(\sigma) \right) P_h x \right\|^2 \, d\sigma \right)^{\frac{1}{2}} \leq C k^{\frac{1+\rho}{2}} \| A_h^{\frac{\ell}{2}} P_h x \| , \tag{3.54}$$

which completes the proof of the case $\rho = 0$. For the case $\rho = 1$ we additionally use (3.13) and Assumption 3.5 which yields

$$\left\| A_h^{\frac{1}{2}} P_h x \right\| = \left\| P_h x \right\|_1 \leq C \| x \|_1$$

for all $x \in \dot{H}^1$. This completes the proof of the Lemma. $\square$

3.6 Spatio-Temporal Approximations of the Stochastic Equation

This section contains our second main result of this chapter, namely the strong error analysis of a spatio-temporal discretization of the stochastic evolution equation (3.1).

Let $k \in (0, 1]$ denote a fixed time step which defines a time grid $t_j = jk$, $j = 0, 1, \ldots, N_k$, with $N_k k \leq T < (N_k + 1)k$. Further, by X_h^j we denote the approximation of the mild solution X to (3.2) at time t_j. A combination of the Galerkin methods together with a linearly implicit Euler–Maruyama scheme results in the recursion

$$X_h^j - X_h^{j-1} + k\left(A_h X_h^j + P_h f(X_h^{j-1})\right) = P_h g(X_h^{j-1})\Delta W^j, \quad \text{for } j = 1,\ldots,N_k,$$

$$X_h^0 = P_h X_0,$$

(3.55)

with the Wiener increments $\Delta W^j := W(t_j) - W(t_{j-1})$, which are $\mathscr{F}_{t_j}$-adapted, U-valued random variables. Consequently, X_h^j is an $\mathscr{F}_{t_j}$-adapted random variable which takes values in S_h.

Our second main result is the analogue of Theorem 3.10.

Theorem 3.14. *Under the assumptions of Sect. 2.3, Assumption 3.1 with $r \in [0,1)$, $p \in [2,\infty)$, and Assumptions 3.3 and 3.5 there exists a constant C, independent of $k,h \in (0,1]$, such that*

$$\|X_h^j - X(t_j)\|_{L^p(\Omega;H)} \le C(h^{1+r} + k^{\frac{1}{2}}), \quad \text{for all } j = 1,\ldots,N_k,$$

where X denotes the mild solution (3.2) to (2.14) and X_h^j is given by (3.55).

As above, we obtain the optimal order of convergence with respect to the spatio-temporal discretization. Since we only use the information of the driving Wiener process which is provided by the increments ΔW^j, it is a well-known fact, see [13], that the maximum order of convergence of the implicit Euler scheme is $\frac{1}{2}$, see also [6,47]. It is possible to overcome this barrier if one considers a Milstein-like scheme as discussed in the recent papers [3,40,48,51].

Proof (of Theorem 3.14). In terms of the rational function $R(kA_h)$, which is introduced in (3.35), we derive the following discrete variation of constants formula for X_h^j

$$X_h^j = R(kA_h)^j P_h X_0 - k \sum_{i=0}^{j-1} R(kA_h)^{j-i} P_h f(t_i, X_h^i)$$

(3.56)

$$+ \sum_{i=0}^{j-1} R(kA_h)^{j-i} P_h g(t_i, X_h^i)\Delta W^{i+1}.$$

By applying (3.2) and (3.56) we get for the error

$$\|X_h^j - X(t_j)\|_{L^p(\Omega;H)}$$

$$\le \left\| R(kA_h)^j P_h X_0 - E(t_j)X_0 \right\|_{L^p(\Omega;H)}$$

$$+ \left\| k \sum_{i=0}^{j-1} R(kA_h)^{j-i} P_h f(t_i, X_h^i) - \int_0^{t_j} E(t_j - \sigma) f(\sigma, X(\sigma))\,d\sigma \right\|_{L^p(\Omega;H)}$$

$$+ \left\| \sum_{i=0}^{j-1} R(kA_h)^{j-i} P_h g(t_i, X_h^i) \Delta W^{i+1} \right.$$

$$\left. - \int_0^{t_j} E(t_j - \sigma) g(\sigma, X(\sigma)) \, dW(\sigma) \right\|_{L^p(\Omega;H)} .$$

(3.57)

The first summand is the error of the fully discrete approximation scheme (3.33) for the homogeneous equation (3.18) but with the initial value being a random variable. By Assumption 2.17 we have that $X_0(\omega) \in \dot{H}^{1+r}$ for P-almost all $\omega \in \Omega$. Thus, the error estimate from [66, Theorem 7.8] yields

$$\left\| R(kA_h)^j P_h X_0 - E(t_j) X_0 \right\|_{L^p(\Omega;H)} \le C \left(h^{1+r} + k^{\frac{1+r}{2}} \right) \left\| A^{\frac{1+r}{2}} X_0 \right\|_{L^p(\Omega;H)} .$$

For the other two summands we introduce two auxiliary processes which are given by

$$f_{kh}(t) := P_h f(t_{j-1}, X_h^{j-1}), \quad \text{if } t \in (t_{j-1}, t_j], \quad j = 1, \ldots, N_k,$$
$$f_{kh}(0) := P_h f(0, X_0),$$

(3.58)

and in the same way

$$g_{kh}(t) := P_h g(t_{j-1}, X_h^{j-1}), \quad \text{if } t \in (t_{j-1}, t_j], \quad j = 1, \ldots, N_k,$$
$$g_{kh}(0) := P_h g(0, X_0).$$

(3.59)

By this definition f_{kh} and g_{kh} are adapted and left-continuous and, hence, predictable processes. Recalling the definition (3.41) of the family of operators $E_{kh}(t)$, $t \ge 0$, we obtain

$$k \sum_{i=0}^{j-1} R(kA_h)^{j-i} P_h f(t_i, X_h^i) = \sum_{i=0}^{j-1} \int_{t_i}^{t_{i+1}} E_{kh}(t_j - \sigma) f_{kh}(\sigma) \, d\sigma$$

$$= \int_0^{t_j} E_{kh}(t_j - \sigma) f_{kh}(\sigma) \, d\sigma$$

(3.60)

and, analogously,

$$\sum_{i=0}^{j-1} R(kA_h)^{j-i} P_h g(t_i, X_h^i) \Delta W^{i+1} = \int_0^{t_j} E_{kh}(t_j - \sigma) g_{kh}(\sigma) \, dW(\sigma). \quad (3.61)$$

By applying (3.60) we have the following estimate of the second summand in (3.57)

$$\left\| k \sum_{i=0}^{j-1} R(kA_h)^{j-i} P_h f(t_i, X_h^i) - \int_0^{t_j} E(t_j - \sigma) f(\sigma, X(\sigma)) \, d\sigma \right\|_{L^p(\Omega;H)}$$

$$= \left\| \int_0^{t_j} \left(E_{kh}(t_j - \sigma) f_{kh}(\sigma) - E(t_j - \sigma) f(\sigma, X(\sigma)) \right) d\sigma \right\|_{L^p(\Omega;H)}$$

$$\leq \left\| \int_0^{t_j} E_{kh}(t_j - \sigma) P_h \left(f_{kh}(\sigma) - f(\sigma, X(\sigma)) \right) d\sigma \right\|_{L^p(\Omega;H)}$$

$$+ \left\| \int_0^{t_j} \left(E_{kh}(t_j - \sigma) P_h - E(t_j - \sigma) \right) \left(f(\sigma, X(\sigma)) - f(t_j, X(t_j)) \right) d\sigma \right\|_{L^p(\Omega;H)}$$

$$+ \left\| \int_0^{t_j} \left(E_{kh}(t_j - \sigma) P_h - E(t_j - \sigma) \right) f(t_j, X(t_j)) \, d\sigma \right\|_{L^p(\Omega;H)}$$

$$=: J_1 + J_2 + J_3.$$

Note, that the terms J_1, J_2, and J_3 are of the same structure as the terms I_1, I_2, and I_3 in the proof of Theorem 3.10. Since with Lemmas 3.12 and 3.13 we have the time discrete analogues of Lemmas 3.8 and 3.9 at our disposal the proof follows the same path.

For the term J_1 we first recall from (3.42) that $\left\| E_{kh}(t) P_h x \right\| \leq C t_i^{-\frac{1}{2}} \left\| x \right\|_{-1}$ for all $t \in [t_{i-1}, t_i)$ and all $x \in \dot{H}^{-1}$. Together with Lemma 2.26 this gives

$$J_1 \leq \int_0^{t_j} \left\| E_{kh}(t_j - \sigma) P_h \left(f_{kh}(\sigma) - f(X(\sigma)) \right) \right\|_{L^p(\Omega;H)} d\sigma$$

$$\leq C \sum_{i=0}^{j-1} \int_{t_i}^{t_{i+1}} t_{j-i}^{-\frac{1}{2}} \left\| f(t_i, X_h^i) - f(\sigma, X(\sigma)) \right\|_{L^p(\Omega;\dot{H}^{-1})} d\sigma$$

$$\leq C k \sum_{i=0}^{j-1} t_{j-i}^{-\frac{1}{2}} \left\| f(t_i, X_h^i) - f(t_i, X(t_i)) \right\|_{L^p(\Omega;\dot{H}^{-1})}$$

$$+ C \sum_{i=0}^{j-1} \int_{t_i}^{t_{i+1}} t_{j-i}^{-\frac{1}{2}} \left\| f(t_i, X(t_i)) - f(\sigma, X(\sigma)) \right\|_{L^p(\Omega;\dot{H}^{-1})} d\sigma$$

$$\leq C k \sum_{i=0}^{j-1} t_{j-i}^{-\frac{1}{2}} \left\| X_h^i - X(t_i) \right\|_{L^p(\Omega;H)}$$

$$+ C \left(1 + \sup_{\sigma \in [0,T]} \left\| X(\sigma) \right\|_{L^p(\Omega;H)} \right) \sum_{i=0}^{j-1} t_{j-i}^{-\frac{1}{2}} \int_{t_i}^{t_{i+1}} (\sigma - t_i)^{\frac{1}{2}} \, d\sigma,$$

where we also used the Hölder continuity (3.4). Since it holds

$$\sum_{i=0}^{j-1} t_{j-i}^{-\frac{1}{2}} \int_{t_i}^{t_{i+1}} (\sigma - t_i)^{\frac{1}{2}} \, d\sigma = \frac{2}{3} Ck^{\frac{3}{2}} \sum_{i=0}^{j-1} t_{j-i}^{-\frac{1}{2}} \leq Ck^{\frac{1}{2}} \int_0^{t_j} \sigma^{-\frac{1}{2}} \, d\sigma \leq CT^{\frac{1}{2}} k^{\frac{1}{2}}.$$

Altogether, J_1 is estimated by

$$J_1 \leq Ck^{\frac{1}{2}} + Ck \sum_{i=0}^{j-1} t_{j-i}^{-\frac{1}{2}} \left\| X_h^i - X(t_i) \right\|_{L^p(\Omega;H)}. \tag{3.62}$$

For the estimate of J_2 we first note that $F_{kh}(t) = E_{kh}(t)P_h - E(t)$ and, hence, we can apply Lemma 3.12(iii) with $\rho = 1 - r$. In the same way as in (3.25) we obtain

$$J_2 \leq C \left(h^{1+r} + k^{\frac{1+r}{2}} \right). \tag{3.63}$$

Likewise, but this time by an application of Lemma 3.13(i) with $\rho = 1 - r$, we proceed with the term J_3 in same way as in (3.26). By also using (3.3) this yields

$$J_3 \leq C \left(h^{1+r} + k^{\frac{1+r}{2}} \right) \left(1 + \sup_{\sigma \in [0,T]} \left\| X(\sigma) \right\|_{L^p(\Omega;H)} \right) \leq C \left(h^{1+r} + k^{\frac{1+r}{2}} \right). \tag{3.64}$$

It remains to estimate the third summand in (3.57) which contains the stochastic integrals. With (3.61) and Proposition 2.12 we get

$$\left\| \sum_{i=0}^{j-1} R(kA_h)^{j-i} P_h g(t_i, X_h^i) \Delta W^{i+1} - \int_0^{t_j} E(t_j - \sigma) g(\sigma, X(\sigma)) \, dW(\sigma) \right\|_{L^p(\Omega;H)}$$

$$= \left\| \int_0^{t_j} E_{kh}(t_j - \sigma) g_{kh}(\sigma) \, dW(\sigma) - \int_0^{t_j} E(t_j - \sigma) g(\sigma, X(\sigma)) \, dW(\sigma) \right\|_{L^p(\Omega;H)}$$

$$\leq C \left(\mathbf{E} \left[\left(\int_0^{t_j} \left\| E_{kh}(t_j - \sigma) g_{kh}(\sigma) - E(t_j - \sigma) g(\sigma, X(\sigma)) \right\|_{L_2^0}^2 \, d\sigma \right)^{\frac{p}{2}} \right] \right)^{\frac{1}{p}}.$$

In the last step Proposition 2.12 is applicable since by our definitions (3.41) and (3.59) the process $[0, t_j] \ni \sigma \mapsto E_{kh}(t_j - \sigma) P_h g_{kh}(\sigma) \in L_2^0$ is adapted and left-continuous and, therefore, predictable.

Next, we use the triangle inequality and obtain

$$\left(\mathbf{E} \left[\left(\int_0^{t_j} \left\| E_{kh}(t_j - \sigma) g_{kh}(\sigma) - E(t_j - \sigma) g(\sigma, X(\sigma)) \right\|_{L_2^0}^2 \, d\sigma \right)^{\frac{p}{2}} \right] \right)^{\frac{1}{p}}$$

$$\leq \left\| \left(\int_0^{t_j} \left\| E_{kh}(t_j - \sigma) P_h \left(g_{kh}(\sigma) - g(\sigma, X(\sigma)) \right) \right\|_{L_2^0}^2 \, d\sigma \right)^{\frac{1}{2}} \right\|_{L^p(\Omega;\mathbb{R})}$$

$$+ \left\|\left(\int_0^{t_j} \left\|F_{kh}(t_j - \sigma)(g(\sigma, X(\sigma)) - g(t_j, X(t_j)))\right\|_{L_2^0}^2 d\sigma\right)^{\frac{1}{2}}\right\|_{L^p(\Omega;\mathbb{R})}$$

$$+ \left\|\left(\int_0^{t_j} \left\|F_{kh}(t_j - \sigma)g(t_j, X(t_j))\right\|_{L_2^0}^2 d\sigma\right)^{\frac{1}{2}}\right\|_{L^p(\Omega;\mathbb{R})}$$

$$=: J_4 + J_5 + J_6.$$

For the estimate of J_4 we use the fact that $\|E_{kh}(t)P_h x\| \le C\|x\|$ for all $x \in H$. Together with Lemma 2.26 we get

$$J_4 \le C\left\|\left(\int_0^{t_j} \|g_{kh}(\sigma) - g(\sigma, X(\sigma))\|^2 d\sigma\right)^{\frac{1}{2}}\right\|_{L^p(\Omega;\mathbb{R})}$$

$$\le C\left(\sum_{i=0}^{j-1} \int_{t_i}^{t_{i+1}} \|g(t_i, X_h^i) - g(\sigma, X(\sigma))\|_{L^p(\Omega;H)}^2 d\sigma\right)^{\frac{1}{2}}$$

$$\le C\left(k \sum_{i=0}^{j-1} \|g(t_i, X_h^i) - g(t_i, X(t_i))\|_{L^p(\Omega;H)}^2\right)^{\frac{1}{2}}$$

$$+ C\left(\sum_{i=0}^{j-1} \int_{t_i}^{t_{i+1}} \|g(t_i, X(t_i)) - g(\sigma, X(\sigma))\|_{L^p(\Omega;H)}^2 d\sigma\right)^{\frac{1}{2}}$$

$$\le C\left(k \sum_{i=0}^{j-1} \|X_h^i - X(t_i)\|_{L^p(\Omega;H)}^2\right)^{\frac{1}{2}}$$

$$+ C\left(1 + \sup_{\sigma \in [0,T]} \|X(\sigma)\|_{L^p(\Omega;H)}\right)\left(\sum_{i=0}^{j-1} \int_{t_i}^{t_{i+1}} (\sigma - t_i) d\sigma\right)^{\frac{1}{2}}$$

Since

$$\sum_{i=0}^{j-1} \int_{t_i}^{t_{i+1}} (\sigma - t_i) d\sigma \le \frac{1}{2}Tk$$

we have shown that

$$J_4 \le Ck^{\frac{1}{2}} + C\left(k \sum_{i=0}^{j-1} \|X_h^i - X(t_i)\|_{L^p(\Omega;H)}^2\right)^{\frac{1}{2}}. \tag{3.65}$$

In the way same as in (3.29), we apply Lemma 3.12(i) with $\mu = 1 + r$ and $\nu = 0$ and we derive the estimate

$$J_5 \le C\left(h^{1+r} + k^{\frac{1+r}{2}}\right), \tag{3.66}$$

where we also used Lemma 2.26 and (3.4). By (3.3) and the same arguments which gave (3.30) we further get

$$J_6 \leq C\left(h^{1+r} + k^{\frac{1+r}{2}}\right). \tag{3.67}$$

To summarize our estimate of (3.57), by (3.62) to (3.67) we have shown that

$$\left\| X_h^j - X(t_j) \right\|_{L^p(\Omega;H)}^2 \leq C\left(h^{1+r} + k^{\frac{1}{2}} + k^{\frac{1+r}{2}}\right)^2$$

$$+ C\left(k \sum_{i=0}^{j-1} t_{j-i}^{-\frac{1}{2}} \left\| X_h^i - X(t_i) \right\|_{L^p(\Omega;H)}\right)^2$$

$$+ Ck \sum_{i=0}^{j-1} \left\| X_h^i - X(t_i) \right\|_{L^p(\Omega;H)}^2.$$

Further, we have

$$k \sum_{i=0}^{j-1} \left\| X_h^i - X(t_i) \right\|_{L^p(\Omega;H)}^2 \leq T^{\frac{1}{2}} k \sum_{i=0}^{j-1} t_{j-i}^{-\frac{1}{2}} \left\| X_h^i - X(t_i) \right\|_{L^p(\Omega;H)}^2,$$

and by Hölder's inequality

$$k \sum_{i=0}^{j-1} t_{j-i}^{-\frac{1}{2}} \left\| X_h^i - X(t_i) \right\|_{L^p(\Omega;H)}$$

$$\leq \left(k \sum_{i=0}^{j-1} t_{j-i}^{-\frac{1}{2}}\right)^{\frac{1}{2}} \left(k \sum_{i=0}^{j-1} t_{j-i}^{-\frac{1}{2}} \left\| X_h^i - X(t_i) \right\|_{L^p(\Omega;H)}^2\right)^{\frac{1}{2}}$$

$$\leq \left(2T^{\frac{1}{2}}\right)^{\frac{1}{2}} \left(k \sum_{i=0}^{j-1} t_{j-i}^{-\frac{1}{2}} \left\| X_h^i - X(t_i) \right\|_{L^p(\Omega;H)}^2\right)^{\frac{1}{2}}.$$

Hence, by setting $\varphi_j = \left\| X_h^j - X(t_j) \right\|_{L^p(\Omega;H)}^2$ we have proven that

$$\varphi_j \leq C\left(h^{1+r} + k^{\frac{1}{2}} + k^{\frac{1+r}{2}}\right)^2 + Ck \sum_{i=0}^{j-1} t_{j-i}^{-\frac{1-r}{2}} \varphi_i.$$

An application of Lemma A.4 completes the proof of the theorem.   □

Chapter 4
A Short Review of the Malliavin Calculus in Hilbert Spaces

So far, the Malliavin calculus has been rarely used in articles covering the numerical analysis of stochastic processes, in particular for SPDEs, and we find it appropriate to provide a gentle and self-contained introduction into this theory.

In order to give a probabilistic proof of Hörmander's hypoelliptic theorem P. Malliavin [55] developed 1976 what he called the stochastic calculus of variations. This method has been expanded by several authors and is nowadays called the Malliavin calculus. For readers which are completely new to this field we refer, for example, to the monograph by D. Nualart [58] which also contains an exhaustive list of references.

In the first two sections we focus on the Malliavin calculus for Hilbert space valued stochastic processes and cylindrical Wiener processes. Most of the presented results are found in [54], which in turn is based on [32]. Since we were not able to locate a reference in the literature, we also give a detailed proof of a chain rule.

4.1 The Derivative Operator

Let $(\Omega, \mathscr{F}, \mathbf{P})$ be a complete probability space with normal filtration $(\mathscr{F}_t)_{t \in [0,T]}$, $t \in [0, T]$. For separable Hilbert spaces $(H, (\cdot, \cdot), \|\cdot\|)$ and $(U, (\cdot, \cdot)_U, \|\cdot\|_U)$ we consider an adapted cylindrical Q-Wiener process W, where the covariance operator satisfies Assumption 2.15. In order to avoid technical issues in the introduction of the Malliavin derivative we assume that $\mathscr{F}$ and the filtration $(\mathscr{F}_t)_{t \in [0,T]}$ are generated by the Wiener process W.

By $C^k(H_1; H_2)$, $k \geq 0$, we denote the set of all continuous mappings $\varphi \colon H_1 \to H_2$, which are k-times continuously Fréchet-differentiable. Since below the letter D is reserved for the Malliavin derivative operator, we write $\varphi' \colon H_1 \to L(H_1, H_2)$ and $\varphi'' \colon H_1 \to L(H_1, L(H_1, H_2))$ for the first and second order Fréchet derivative.

R. Kruse, *Strong and Weak Approximation of Semilinear Stochastic Evolution Equations*, Lecture Notes in Mathematics 2093, DOI 10.1007/978-3-319-02231-4_4, © Springer International Publishing Switzerland 2014

Further for $\varphi \in C^2(H; \mathbb{R})$, by using the representation theorem, we understand the first derivative $\varphi'(h)$ at $h \in H$ as an element in H. Similarly, the second derivative $\varphi''(h)$ at $h \in H$ is identified with a self-adjoint linear operator in $L(H)$, that is

$$\left[\frac{d^2}{dh^2} \varphi(h) \right](h_1, h_2) = \left(\varphi''(h)h_1, h_2 \right), \quad \text{for all } h, h_1, h_2 \in H.$$

The set $C_p^k(H_1; H_2)$ contains all mappings $\varphi \in C^k(H_1; H_2)$, where φ and all its derivatives are at most polynomially growing. Similarly, the set $C_b^k(H_1; H_2)$ consists of all mappings $\varphi \in C^k(H_1; H_2)$ such that φ and all its derivatives are bounded.

For a deterministic mapping $\Phi \in L^2([0, T]; L_2(U_0, \mathbb{R}))$ let us define

$$W(\Phi) := \int_0^T \Phi(t) \, dW(t),$$

where the integral on the right hand side is the usual Itô-integral with respect to the cylindrical Wiener process W as in Sect. 2.2.

We now introduce the first important operator of the Malliavin calculus, the so called *Malliavin derivative*. As in [32,54] and [59, Sect. 2.1] we define $\mathscr{S}(H)$ to be the set of all smooth and cylindrical random variables F of the form

$$F = \sum_{j=1}^n f_j(W(\Phi_1), \dots, W(\Phi_m))h_j, \tag{4.1}$$

where $m, n \geq 1$, $\Phi_i \in L^2([0, T]; L_2(U_0, \mathbb{R}))$, $1 \leq i \leq m$, and $h_j \in H$, $f_j \in C_p^1(\mathbb{R}^m; \mathbb{R})$ for $1 \leq j \leq n$. In particular, from Proposition 2.12 it directly follows that $\mathscr{S}(H) \subset L^p(\Omega; H)$ for all $p \geq 1$.

Remark 4.1. In [32, 54] and [59, Sect. 2.1] the authors require that $f_j \in C_b^\infty(\mathbb{R}^m; \mathbb{R})$ or $f_j \in C_p^\infty(\mathbb{R}^m; \mathbb{R})$, $j = 1, \dots, n$. However, we decided to enlarge the set $\mathscr{S}(H)$ by decreasing the regularity requirement on f_j. By doing this, we avoid some complications in the proof of the chain rule (see Lemma 4.7 below).

In the same way as in [32,54] we define the derivative DF of $F \in \mathscr{S}(H)$ by

$$DF(t) := \sum_{j=1}^n \sum_{i=1}^m \frac{\partial}{\partial x_i} f_j(W(\Phi_1), \dots, W(\Phi_m))h_j \otimes \Phi_i(t), \quad \text{for } t \in [0, T],$$

$$\tag{4.2}$$

where $h_j \otimes \Phi_i(t)$ denotes the tensor product of $h_j \in H$ and $\Phi_i(t) \in L_2(U_0, \mathbb{R})$, that is, for $1 \leq i \leq m$ and $1 \leq j \leq n$,

$$(h_j \otimes \Phi_i(t))(u) = (\Phi_i(t)(u))h_j \in H \quad \text{for all } u \in U_0, \ t \in [0, T]. \tag{4.3}$$

Since for an arbitrary orthonormal basis $(e_\ell)_{\ell \geq 1}$ of U it holds that

$$\|h_j \otimes \Phi_i(t)\|_{L_2^0}^2 = \sum_{\ell=1}^{\infty} \left\|(\Phi(t)(Q^{\frac{1}{2}}e_\ell))h_j\right\|^2 = \|h_j\|^2 \|\Phi(t)\|_{L_2(U_0,\mathbb{R})}^2, \qquad (4.4)$$

we obtain $h_j \otimes \Phi_i(t) \in L_2^0$ for almost every $t \in [0, T]$. Hence, DF may be interpreted as a smooth and cylindrical random variable which maps into the Hilbert space $L^2([0, T]; L_2^0)$ provided that the appearing partial derivatives of f_j, $j = 1, \ldots, n$, are smooth enough.

Proposition 4.2. *The operator $D: \mathscr{S}(H) \subset L^p(\Omega; H) \to L^p(\Omega; L^2([0, T]; L_2^0))$ is well-defined. In particular, DF does not depend on the representation of $F \in \mathscr{S}(H)$.*

Proof. Consider $F \in \mathscr{S}(H)$ with representation

$$F = \sum_{j=1}^{n} f_j(W(\Phi_1), \ldots, W(\Phi_m))h_j.$$

First, by linearity of the operator D, we may assume without loss of generality that $(h_j)_{j=1,\ldots,n}$ forms an orthonormal basis of $\mathrm{span}\{h_1, \ldots, h_n\} \subset H$.

From this it follows, again without loss of generality, that it is enough to consider $F \in \mathscr{S}(H)$ which can be represented by

$$F = f(W(\Phi_1), \ldots, W(\Phi_m))h$$

with $\|h\| = 1$.

Next, let $(\Psi_i)_{i=1,\ldots,k}$, $k \leq m$, denote an orthonormal basis of $\mathrm{span}\{\Phi_1, \ldots, \Phi_m\} \subset L^2([0, T]; L_2(U_0; \mathbb{R}))$. Since $W(\cdot)$ is linear, we have

$$W(\Phi_j) = \sum_{i=1}^{k} (\Phi_j, \Psi_i)_{L^2([0,T];L_2^0)} W(\Psi_i)$$

and we define $g \in C_p^1(\mathbb{R}^k; \mathbb{R})$ by

$$g(x_1, \ldots, x_k) := f\left(\sum_{i=1}^{k} (\Phi_1, \Psi_i)_{L^2([0,T];L_2^0)} x_i, \ldots, \sum_{i=1}^{k} (\Phi_m, \Psi_i)_{L^2([0,T];L_2^0)} x_i\right).$$

This provides us with a new representation of F, namely

$$F = g(W(\Psi_1), \ldots, W(\Psi_k))h.$$

Then, it holds that

$$\frac{\partial}{\partial x_i} g = \sum_{j=1}^{m} (\Phi_j, \Psi_i)_{L^2([0,T];L_2^0)} \frac{\partial}{\partial x_j} f$$

and, consequently,

$$DF = \sum_{i=1}^{k} \frac{\partial}{\partial x_i} g(W(\Psi_1), \ldots, W(\Psi_k)) h \otimes \Psi_i$$

$$= \sum_{i=1}^{k} \sum_{j=1}^{m} (\Phi_j, \Psi_i)_{L^2([0,T];L_2^0)} \frac{\partial}{\partial x_j} f(W(\Phi_1), \ldots, W(\Phi_m)) h \otimes \Psi_i$$

$$= \sum_{j=1}^{m} \frac{\partial}{\partial x_j} f(W(\Phi_1), \ldots, W(\Phi_m)) \Big(h \otimes \sum_{i=1}^{k} (\Phi_j, \Psi_i)_{L^2([0,T];L_2^0)} \Psi_i \Big)$$

$$= \sum_{j=1}^{m} \frac{\partial}{\partial x_j} f(W(\Phi_1), \ldots, W(\Phi_m)) h \otimes \Phi_j,$$

that is, DF coincides for both representations of F.

In the general case, where $F \in \mathscr{S}(H)$ has two representations of the form

$$F = f(W(\Phi_1), \ldots, W(\Phi_m)) h = \tilde{f}(W(\Upsilon_1), \ldots, W(\Upsilon_\ell)) h$$

we come to the same conclusion if we apply the given arguments to an orthonormal basis of $\mathrm{span}\{\Phi_1, \ldots, \Phi_m, \Upsilon_1, \ldots, \Upsilon_\ell\} \subset L^2([0,T]; L_2^0)$. Therefore, the definition of DF is indeed independent of the particular representation of $F \in \mathscr{S}(H)$.   □

Following in the footsteps of [59, Sect. 2] the next result may be interpreted as an integration by parts formula (cf. [59, Prop. 2.3]).

Proposition 4.3. *Suppose that $F \in \mathscr{S}(H)$ and $\Psi \in L^2([0,T]; L_2^0)$. Then*

$$\mathbf{E}\Big[\Big(F, \int_0^T \Psi(t)\, dW(t)\Big)\Big] = \mathbf{E}\big[(DF, \Psi)_{L^2([0,T];L_2^0)}\big].$$

Proof. As in the proof of Proposition 4.2 we choose a representation

$$F = \sum_{j=1}^{n} f_j(W(\Phi_1), \ldots, W(\Phi_m)) h_j$$

such that $(h_j)_{j=1,\ldots,n}$ and $(\Phi_i)_{i=1,\ldots,m}$ form orthonormal systems in H and $L^2([0,T]; L_2(U_0, \mathbb{R}))$, respectively.

Then, by Proposition 2.11, we obtain

$$\mathbf{E}\Big[\Big(F, \int_0^T \Psi(t)\,dW(t)\Big)\Big] = \sum_{j=1}^n \mathbf{E}\Big[f_j(W(\Phi_1),\dots,W(\Phi_m))\int_0^T \big(h_j, \Psi(t)\,dW(t)\big)\Big]$$

$$= \sum_{j=1}^n \mathbf{E}\Big[f_j(W(\Phi_1),\dots,W(\Phi_m))\int_0^T \Psi_{h_j}(t)\,dW(t)\Big]$$

$$= \sum_{j=1}^n \mathbf{E}\big[f_j(W(\Phi_1),\dots,W(\Phi_m))W(\Psi_{h_j})\big],$$

where $\Psi_{h_j} \in L^2([0,T]; L_2(U_0,\mathbb{R}))$ is given by

$$\Psi_{h_j}(t)(u) := (h_j, \Psi(t)u), \quad \text{for } u \in U_0.$$

Now, we make use of the following decomposition of Ψ_{h_j}

$$\Psi_{h_j} = \sum_{i=1}^m (\Psi_{h_j}, \Phi_i)_{L^2([0,T];L_2(U_0,\mathbb{R}))}\Phi_i + \Psi_{h_j}^\perp,$$

where $\Psi_{h_j}^\perp \in L^2([0,T]; L_2(U_0,\mathbb{R}))$ is the uniquely determined residual in the orthogonal complement of $\mathrm{span}\{\Phi_1,\dots,\Phi_m\}$. Hence, we have

$$\mathbf{E}\Big[\Big(F, \int_0^T \Psi(t)\,dW(t)\Big)\Big]$$

$$= \sum_{j=1}^n \sum_{i=1}^m (\Psi_{h_j}, \Phi_i)_{L^2([0,T];L_2(U_0,\mathbb{R}))}\mathbf{E}\big[f_j(W(\Phi_1),\dots,W(\Phi_m))W(\Phi_i)\big]$$

$$+ \sum_{j=1}^n \mathbf{E}\big[f_j(W(\Phi_1),\dots,W(\Phi_m))W(\Psi_{h_j}^\perp)\big].$$

Since the integrands Φ_i, $i = 1,\dots,m$, and $\Psi_{h_j}^\perp$ are deterministic, it follows from the construction of the Itô-integral that $(W(\Phi_1),\dots,W(\Phi_m),W(\Psi_{h_j}^\perp))$ is an $\mathbb{R}^{m+1}$-valued random vector, whose components are pairwise uncorrelated, centered Gaussian random variables. Therefore, for every $j = 1,\dots,n$, it holds that

$$\mathbf{E}\big[f_j(W(\Phi_1),\dots,W(\Phi_m))W(\Psi_{h_j}^\perp)\big]$$

$$= \mathbf{E}\big[f_j(W(\Phi_1),\dots,W(\Phi_m))\big]\mathbf{E}\big[W(\Psi_{h_j}^\perp)\big] = 0.$$

Further, since $E[W(\Phi_i)^2] = 1, i = 1, \ldots, m$, the probability density function of the vector $(W(\Phi_1), \ldots, W(\Phi_m))$ is given by

$$\varphi(x_1, \ldots, x_m) = (2\pi)^{-\frac{m}{2}} \exp\left(-\frac{1}{2}\sum_{i=1}^{m} x_i^2\right)$$

and we get by the real-valued integration by parts formula

$$E\big[f_j(W(\Phi_1), \ldots, W(\Phi_m))W(\Phi_i)\big]$$

$$= \int_{\mathbb{R}^m} f_j(x_1, \ldots, x_m)\, x_i\, \varphi(x_1, \ldots, x_m)\, d(x_1, \ldots, x_m)$$

$$= \int_{\mathbb{R}^m} \frac{\partial}{\partial x_i} f_j(x_1, \ldots, x_m)\varphi(x_1, \ldots, x_m)\, d(x_1, \ldots, x_m)$$

$$= E\Big[\frac{\partial}{\partial x_i} f_j(W(\Phi_1), \ldots, W(\Phi_m))\Big].$$

Altogether, this yields

$$E\Big[\Big(F, \int_0^T \Psi(t)\, dW(t)\Big)\Big]$$

$$= \sum_{j=1}^{n}\sum_{i=1}^{m} (\Psi_{h_j}, \Phi_i)_{L^2([0,T];L_2(U_0,\mathbb{R}))} E\Big[\frac{\partial}{\partial x_i} f_j(W(\Phi_1), \ldots, W(\Phi_m))\Big]$$

$$= \sum_{j=1}^{n}\sum_{i=1}^{m} (h_j \otimes \Phi_i, \Psi)_{L^2([0,T];L_2^0)} E\Big[\frac{\partial}{\partial x_i} f_j(W(\Phi_1), \ldots, W(\Phi_m))\Big]$$

$$= E\big[(DF, \Psi)_{L^2([0,T];L_2^0)}\big]$$

and the proof is complete. $\square$

Our next proposition gives a comparable result to [59, Prop. 2.5].

Proposition 4.4. *The operator* $D: \mathscr{S}(H) \subset L^p(\Omega; H) \to L^p(\Omega; L^2([0,T]; L_2^0))$ *is closable for every* $p > 1$.

Proof. Let $(F_\ell)_{\ell \geq 1}$ be a sequence in $\mathscr{S}(H)$ such that

$$F_\ell \to 0 \qquad \text{in } L^p(\Omega; H) \text{ as } \ell \to \infty,$$

$$DF_\ell \to \Upsilon \qquad \text{in } L^p(\Omega; L^2([0,T]; L_2^0)) \text{ as } \ell \to \infty.$$

In order to prove that D is closable we have to show that $\Upsilon = 0$. For this, we consider a smooth random variable $G: \Omega \to L^2([0,T]; L_2^0)$ of the form

$$G = g(W(\Phi_1), \ldots, W(\Phi_m))\Psi$$

with $g \in C_b^1(\mathbb{R}^m)$ and $\Psi \in L^2([0, T]; L_2^0)$, $\Phi_i \in L^2([0, T]; L_2(U_0, \mathbb{R}))$. We remark that in this situation $g(W(\Phi_1), \ldots, W(\Phi_m))F_\ell$ is a smooth and cylindrical random variable and the product rule yields

$$D(g(W(\Phi_1), \ldots, W(\Phi_m))F_\ell) = F_\ell \otimes Dg + g(W(\Phi_1), \ldots, W(\Phi_m))DF_\ell. \tag{4.5}$$

Together with Proposition 4.3 we get (compare also with [58, Lem. 1.2.2] and [59, Prop. 2.4])

$$\mathbf{E}\big[(\Upsilon, G)_{L^2([0,T];L_2^0)}\big] = \lim_{\ell \to \infty} \mathbf{E}\big[(DF_\ell, g(W(\Phi_1), \ldots, W(\Phi_m))\Psi)_{L^2([0,T];L_2^0)}\big]$$

$$= \lim_{\ell \to \infty} \mathbf{E}\Big[\Big(g(W(\Phi_1), \ldots, W(\Phi_m))F_\ell, \int_0^T \Psi(t)\,\mathrm{d}W(t)\Big)$$

$$- \big(F_\ell \otimes Dg, \Psi\big)_{L^2([0,T];L_2^0)}\Big].$$

Finally, we get by an application of Hölder's inequality

$$\Big|\mathbf{E}\Big[\Big(g(W(\Phi_1), \ldots, W(\Phi_m))F_\ell, \int_0^T \Psi(t)\,\mathrm{d}W(t)\Big)\Big]\Big|$$

$$\leq \mathbf{E}\Big[\big\|g(W(\Phi_1), \ldots, W(\Phi_m))F_\ell\big\| \Big\|\int_0^T \Psi(t)\,\mathrm{d}W(t)\Big\|\Big]$$

$$\leq \big\|g(W(\Phi_1), \ldots, W(\Phi_m))F_\ell\big\|_{L^q(\Omega;H)} \Big\|\int_0^T \Psi(t)\,\mathrm{d}W(t)\Big\|_{L^{q'}(\Omega;H)}$$

with $q = \min(2, p)$ and q' its adjoint exponent such that $1 = \frac{1}{q} + \frac{1}{q'}$. Since $2 \leq q' < \infty$, we have by Proposition 2.12 that

$$\Big\|\int_0^T \Psi(t)\,\mathrm{d}W(t)\Big\|_{L^{q'}(\Omega;H)} \leq C(q')\|\Psi\|_{L^2([0,T];L_2^0)} < \infty. \tag{4.6}$$

Further, since $g \in C_b^1(\mathbb{R}^m; \mathbb{R})$, $F_\ell \to 0$ in $L^p(\Omega; H)$ as $\ell \to \infty$ and $q \leq p$, it also holds that

$$\lim_{\ell \to \infty} \big\|g(W(\Phi_1), \ldots, W(\Phi_m))F_\ell\big\|_{L^q(\Omega;H)} = 0.$$

Similarly, by (4.4) it is true that

$$\lim_{\ell \to \infty} \big|\mathbf{E}\big[(F_\ell \otimes Dg, \Psi)_{L^2([0,T];L_2^0)}\big]\big|$$

$$\leq \|\Psi\|_{L^2([0,T];L_2^0)} \lim_{\ell \to \infty} \mathbf{E}\big[\|F_\ell\|\|Dg\|_{L^2([0,T];L_2(U_0,\mathbb{R}))}\big]$$

$$\leq \|\Psi\|_{L^2([0,T];L_2^0)} \|Dg\|_{L^{p'}(\Omega;L^2([0,T];L_2(U_0,\mathbb{R})))} \lim_{\ell \to \infty} \|F_\ell\|_{L^p(\Omega;H)} = 0.$$

Therefore,

$$\mathbf{E}\big[(\Upsilon, G)_{L^2([0,T];L_2^0)}\big] = 0.$$

This proves that $\Upsilon = 0$ since the linear span of all G of the above form is dense in $L^{p'}(\Omega; L^2([0, T], L_2^0))$.   □

Remark 4.5. In Proposition 4.4 we excluded the case $p = 1$ which is due to the fact that (4.6) does not hold for $q' = \infty$. In the real-valued situation it is indicated in [58, Prop. 1.2.1] how to treat this case. A similar approach may also work in our Hilbert space valued setting. However, since for our purposes the case $p = 1$ plays no important role we omit this detail.

By following the usual procedure in [58, Chap. 1.2] and [59, Sect. 2] we introduce $\mathbb{D}^{1,p}(H)$ for $1 < p < \infty$ as the closure of $\mathscr{S}(H)$ in $L^p(\Omega; H)$ with respect to the seminorm

$$\|F\|_{\mathbb{D}^{1,p}(H)} := \Big(\mathbf{E}\big[\|F\|^p\big] + \mathbf{E}\big[\|DF\|^p_{L^2([0,T];L_2^0)}\big]\Big)^{\frac{1}{p}} \tag{4.7}$$

and we obtain a well-defined extension of the operator $D: \mathbb{D}^{1,p}(H) \subset L^p(\Omega; H) \to L^p(\Omega; L^2([0, T]; L_2^0))$. In the following we often consider $\mathbb{D}^{1,\infty}(H) := \cap_{p>1}\mathbb{D}^{1,p}(H)$. Note that $\mathscr{S}(H) \subset \mathbb{D}^{1,\infty}(H)$.

We also remark that, for $F \in \mathbb{D}^{1,p}(H)$, DF is a stochastic process with values in L_2^0. Sometimes we want to stress how $DF(t)$ acts on elements of U_0 and in this case we also write

$$DF(t; u) := DF(t)u, \quad t \in [0, T], \ u \in U_0. \tag{4.8}$$

Thus, for fixed $t \in [0, T]$, $u \in U_0$, $DF(t; u)$ is an H-valued random variable.

From the construction of D it is also obvious how to define iterated derivatives D^k. For this we introduce the set $\mathscr{S}^k(H)$ of smooth and cylindrical random variables with increased regularity, that is, for every $F \in \mathscr{S}^k(H)$ there exists a representation of the form (4.1) with $f_j \in C_p^k(\mathbb{R}^m; \mathbb{R})$.

The domain of D^k is denoted by $\mathbb{D}^{k,p}(H)$ with norm

$$\|F\|_{\mathbb{D}^{k,p}(H)} := \Big(\mathbf{E}\big[\|F\|^p\big] + \sum_{j=1}^{k} \mathbf{E}\big[\|D^j F\|^p_{L^2([0,T]^j; H \otimes U_0^{\otimes j})}\big]\Big)^{\frac{1}{p}},$$

where

$$H \otimes U_0^{\otimes j} \simeq L_2(U_0, \underbrace{\ldots}_{j \text{ times}}, L_2(U_0, H)).$$

For example, the second derivative $D^2 F$ takes values in the space

$$L^2([0, T]; L_2(U_0, L^2([0, T]; L_2^0))) \simeq L^2([0, T]^2; L_2(U_0, L_2^0)).$$

Example 4.6. In this example we show that the Wiener process W lies in the domain of the Malliavin derivative. For simplicity we set $H = U$ and we assume that the covariance operator $Q: H \to H$ is trace class, that is $\mathrm{Tr}(Q) < \infty$.

For $t \in [0, T]$ the Wiener process $W(t)$ can be represented by Proposition 2.3

$$W(t) = \sum_{k=1}^{\infty} \sqrt{\mu_k} \beta_k(t) e_k,$$

where $(\beta_k)_{k \in \mathbb{N}}$ are independent real-valued Brownian motions and $(e_k)_{k \in \mathbb{N}}$ denotes an orthonormal basis of H such that $Q e_k = \mu_k e_k$ with $\mu_k \geq 0$ for all $k \in \mathbb{N}$.

In this situation we define $\Phi_{k,t}: [0, T] \times \Omega \to L_2(U_0, \mathbb{R})$, $t \in [0, T]$, $k \in \mathbb{N}$, by

$$\Phi_{k,t}(\sigma) = 1_{[0,t]}(\sigma)(e_k, \cdot), \quad \text{for } \sigma \in [0, T].$$

Thus, it holds $W(\Phi_{k,t}) = \sqrt{\mu_k} \beta_k(t)$. Now, for $N \in \mathbb{N}$ we consider

$$W^N(t) := \sum_{k=1}^{N} \sqrt{\mu_k} \beta_k(t) e_k = \sum_{k=1}^{N} W(\Phi_{k,t}) e_k.$$

Clearly, $W^N(t) \in \mathscr{S}(H)$ and $W^N(t) \to W(t)$ with respect to the norm in $L^2(\Omega; H)$. Further, we have

$$[DW^N(t)](\sigma) = \sum_{k=1}^{N} e_k \otimes \Phi_{k,t}(\sigma) = \sum_{k=1}^{N} 1_{[0,t]}(\sigma) e_k \otimes e_k.$$

Since

$$\|DW^N(t)\|_{L^2([0,T];L_2^0)}^2 = \int_0^t \sum_{k=1}^{N} \|e_k \otimes e_k\|_{L_2^0}^2 \, d\sigma = t \sum_{k=1}^{N} \mu_k \leq t \, \mathrm{Tr}(Q),$$

we obtain that

$$[DW^N(t)](\sigma) \to \sum_{k=1}^{\infty} 1_{[0,t]}(\sigma) e_k \otimes e_k =: G(\sigma) \quad \text{as } N \to \infty$$

with respect to the norm in $L^2([0, T]; L_2^0)$. Since D is a closed operator and $G \in L^2([0, T]; L_2^0)$, we obtain $W(t) \in \mathbb{D}^{1,2}(H)$ for a fixed $t \in [0, T]$ and $DW(t) = G$.

The next result is a chain rule for the derivative operator, which is a variant of [58, Prop. 1.2.3].

Lemma 4.7. *Let* $\varphi: H \to \tilde{H}$ *be a continuously Fréchet-differentiable mapping, where* $(\tilde{H}, (\cdot, \cdot)_{\tilde{H}}, \|\cdot\|_{\tilde{H}})$ *is an arbitrary separable Hilbert space.*
Assume that there exists a $q \geq 0$ *and a constant* $C > 0$ *such that*

$$\|\varphi(h)\|_{\tilde{H}} \leq C(1 + \|h\|^{1+q}), \quad \|\varphi'(h)\|_{L(H,\tilde{H})} \leq C(1 + \|h\|^q) \quad \text{for all } h \in H. \tag{4.9}$$

Then, for all $p > 1$ *and* $F \in \mathbb{D}^{1,(q+1)p}(H)$ *it holds that* $\varphi(F) \in \mathbb{D}^{1,p}(\tilde{H})$ *and*

$$D\varphi(F) = \varphi'(F)DF. \tag{4.10}$$

In particular, if $F \in \mathbb{D}^{1,\infty}(H)$ *then* $\varphi(F) \in \mathbb{D}^{1,\infty}(\tilde{H})$.

Proof. Since $F \in \mathbb{D}^{1,(1+q)p}(H)$ there exists a sequence of smooth and cylindrical random variables $(F_\ell)_{\ell \in \mathbb{N}} \subset \mathscr{S}(H)$ such that

$$\|F - F_\ell\|_{\mathbb{D}^{1,(1+q)p}} \to 0 \quad \text{as } \ell \to \infty.$$

Further, we consider the following approximations $(\varphi_\nu)_{\nu \in \mathbb{N}}$ of φ, which are given by

$$\varphi_\nu(h) = \sum_{\eta=1}^{\nu} (\varphi(h), \tilde{e}_\eta)_{\tilde{H}} \tilde{e}_\eta, \quad \text{for all } h \in H,$$

where $(\tilde{e}_\eta)_{\eta \in \mathbb{N}}$ denotes an arbitrary orthonormal basis of $\tilde{H}$.

For an arbitrary $\ell \in \mathbb{N}$ assume that F_ℓ has the representation (4.1) such that $(h_j)_{j=1,\dots,n}$ and $(\Phi_i)_{i=1,\dots,m}$ form orthonormal systems in the separable Hilbert spaces H and $L^2([0,T]; L_2(U_0, \mathbb{R}))$, respectively.

Now, for every $\nu \in \mathbb{N}$ we get that

$$\varphi_\nu(F_\ell) = \sum_{\eta=1}^{\nu} (\varphi(F_\ell), \tilde{e}_\eta)_{\tilde{H}} \tilde{e}_\eta = \sum_{\eta=1}^{\nu} b_\eta(W(\Phi_1), \dots, W(\Phi_m)) \tilde{e}_\eta,$$

where the function $b_\eta: \mathbb{R}^m \to \mathbb{R}$, $\eta \in \mathbb{N}$, is given by

$$b_\eta(x_1, \dots, x_m) := \left(\varphi\left(\sum_{j=1}^{n} f_j(x_1, \dots, x_m) h_j \right), \tilde{e}_\eta \right)_{\tilde{H}}. \tag{4.11}$$

Since $b_\eta \in C_p^1(\mathbb{R}^m; \mathbb{R})$, we obtain $\varphi_\nu(F_\ell) \in \mathscr{S}(\tilde{H}) \subset \mathbb{D}^{1,p}(\tilde{H})$. Further, by using (4.2) and

$$\frac{\partial}{\partial x_i} b_\eta(W(\Phi_1), \dots, W(\Phi_m)) = \sum_{j=1}^{n} \frac{\partial}{\partial x_i} f_j(W(\Phi_1), \dots, W(\Phi_m))(\varphi'(F_\ell)h_j, \tilde{e}_\eta)_{\tilde{H}}$$

as well as

$$\varphi_\nu'(h)h_j = \sum_{\eta=1}^{\nu} (\varphi'(h)h_j, \tilde{e}_\eta)_{\tilde{H}} \tilde{e}_\eta \tag{4.12}$$

we derive, for all $t \in [0, T]$,

$$
\begin{aligned}
D\varphi_\nu(F_\ell)(t) &= \sum_{\eta=1}^{\nu}\sum_{i=1}^{m} \frac{\partial}{\partial x_i} b_\eta(W(\Phi_1), \dots, W(\Phi_m))\tilde{e}_\eta \otimes \Phi_i(t) \\
&= \sum_{\eta=1}^{\nu}\sum_{j=1}^{n}\sum_{i=1}^{m} \frac{\partial}{\partial x_i} f_j(W(\Phi_1), \dots, W(\Phi_m))(\varphi'(F_\ell)h_j, \tilde{e}_\eta)_{\tilde{H}} \tilde{e}_\eta \otimes \Phi_i(t) \\
&= \sum_{j=1}^{n}\sum_{i=1}^{m} \frac{\partial}{\partial x_i} f_j(W(\Phi_1), \dots, W(\Phi_m))(\varphi_\nu'(F_\ell)h_j) \otimes \Phi_i(t) \\
&= \varphi_\nu'(F_\ell)DF_\ell(t).
\end{aligned}
$$

Altogether, this proves (4.10) for $\varphi_\nu(F_\ell)$ for all $\nu, \ell \in \mathbb{N}$. In order to prove (4.10) for general $\varphi \in C_p^1(H; \tilde{H})$ with (4.9) and $F \in \mathbb{D}^{1,(1+q)p}(H)$, we first note that

$$\|\varphi(F) - \varphi_\nu(F_\ell)\|_{L^p(\Omega;\tilde{H})} \leq \|\varphi(F) - \varphi(F_\ell)\|_{L^p(\Omega;\tilde{H})} + \|\varphi(F_\ell) - \varphi_\nu(F_\ell)\|_{L^p(\Omega;\tilde{H})}. \tag{4.13}$$

By the mean value theorem and (4.9) it holds that

$$\|\varphi(h_1) - \varphi(h_2)\| \leq C\left(1 + \|h_1\|^q + \|h_2\|^q\right)\|h_1 - h_2\|, \quad \text{for all } h_1, h_2 \in H.$$

Thus,

$$
\begin{aligned}
&\|\varphi(F) - \varphi(F_\ell)\|_{L^p(\Omega;\tilde{H})} \\
&\quad \leq C\left(\mathbf{E}\left[\left(1 + \|F\|^q + \|F_\ell\|^q\right)^p \|F - F_\ell\|^p\right]\right)^{\frac{1}{p}} \\
&\quad \leq C\left(1 + \|F\|_{L^{(1+q)p}(\Omega;H)}^q + \|F_\ell\|_{L^{(1+q)p}(\Omega;H)}^q\right)\|F - F_\ell\|_{L^{(1+q)p}(\Omega;H)},
\end{aligned}
$$

where we applied Hölder's inequality with exponents $r = \frac{(1+q)p}{q}$ and $r' = (1+q)p$ such that $\frac{1}{p} = \frac{1}{r} + \frac{1}{r'}$ (note that the case $r = \infty$ causes no additional difficulties). Hence the first summand in (4.13) is getting small as $\ell \to \infty$.

This is also true for the second summand: By the construction of φ_ν we have that $\varphi_\nu(h) \to \varphi(h)$ for every h in the norm in $\tilde{H}$ as $\nu \to \infty$. Thus, $\varphi_\nu(F_\ell(\omega)) \to \varphi(F_\ell(\omega))$ for almost all $\omega \in \Omega$ as $\nu \to \infty$. A dominating function is obtained by $\|\varphi(F_\ell(\omega)) - \varphi_\nu(F_\ell(\omega))\|_{\tilde{H}} \leq \|\varphi(F_\ell(\omega))\|_{\tilde{H}}$ and we get the desired result from Lebesgue's dominated convergence theorem.

Therefore, it remains to prove that

$$\left\| D\varphi_\nu(F_\ell) - \varphi'(F)DF \right\|_{L^p(\Omega;L_2(U_0,\tilde{H}))} \to 0 \quad \text{as } \ell, \nu \to \infty.$$

By using (4.10) for $\varphi_\nu(F_\ell)$ we get

$$\left\| D\varphi_\nu(F_\ell) - \varphi'(F)DF \right\|_{L^p(\Omega;L^2([0,T];L_2(U_0,\tilde{H})))}$$

$$\leq \left\| \varphi'_\nu(F_\ell)DF_\ell - \varphi'_\nu(F_\ell)DF \right\|_{L^p(\Omega;L^2([0,T];L_2(U_0,\tilde{H})))}$$

$$+ \left\| \varphi'_\nu(F_\ell)DF - \varphi'_\nu(F)DF \right\|_{L^p(\Omega;L^2([0,T];L_2(U_0,\tilde{H})))} \tag{4.14}$$

$$+ \left\| \varphi'_\nu(F)DF - \varphi'(F)DF \right\|_{L^p(\Omega;L^2([0,T];L_2(U_0,\tilde{H})))}.$$

We discuss these three summands separately. By (4.12) and (4.9) it holds that

$$\left\| \varphi'_\nu(h_1)h_2 \right\|_{\tilde{H}} = \left(\sum_{\eta=1}^{\nu} \left| (\varphi'(h_1)h_2, \tilde{e}_\eta)_{\tilde{H}} \right|^2 \right)^{\frac{1}{2}}$$

$$\leq \left(\sum_{\eta=1}^{\infty} \left| (\varphi'(h_1)h_2, \tilde{e}_\eta)_{\tilde{H}} \right|^2 \right)^{\frac{1}{2}}$$

$$= \left\| \varphi'(h_1)h_2 \right\|_{\tilde{H}} \leq \left\| \varphi'(h_1) \right\|_{L(H,\tilde{H})} \|h_2\| \leq C(1 + \|h_1\|^q)\|h_2\| \tag{4.15}$$

for all $h_1, h_2 \in H$, where the constant C is independent of $\nu \in \mathbb{N}$. Then, an application of Hölder's inequality with the same exponents as above yields

$$\left\| \varphi'_\nu(F_\ell)DF_\ell - \varphi'_\nu(F_\ell)DF \right\|_{L^p(\Omega;L^2([0,T];L_2(U_0,\tilde{H})))}$$

$$\leq C\left(1 + \|F_\ell\|_{L^{(1+q)p}(\Omega;H)}^q\right)\|DF_\ell - DF\|_{L^{(1+q)p}(\Omega;L^2([0,T];L_2^0))} \to 0 \quad \text{as } \ell \to \infty.$$

Next, we deal with the second summand: From (4.12) and the continuity of φ' it follows that also φ'_ν is continuous for every $\nu \in \mathbb{N}$. Further, by possibly considering a suitable subsequence of $(F_\ell)_{\ell \in \mathbb{N}}$ we may assume that

$$F_\ell(\omega) \to F(\omega) \quad \text{as } \ell \to \infty \quad \text{for almost all } \omega \in \Omega.$$

From this it follows that

$$\lim_{\ell \to \infty} \left\| (\varphi'_\nu(F_\ell(\omega)) - \varphi'_\nu(F(\omega)))DF(t,\omega)e_\mu \right\|_{\tilde{H}} = 0,$$

for almost all $\omega \in \Omega$, $t \in [0,T]$, and all $\mu \in \mathbb{N}$, where $(e_\mu)_{\mu \in \mathbb{N}}$ denotes an orthonormal basis of U_0 as in (2.8). Further, by using our bound on φ'_ν from (4.15), a dominating function is obtained by

$$\|(\varphi'_\nu(F_\ell(\omega)) - \varphi'_\nu(F(\omega)))DF(t,\omega)e_\mu\|_{\tilde{H}}$$
$$\leq C(1 + \|F_\ell(\omega)\|^q + \|F(\omega)\|^q)\|DF(t,\omega)e_\mu\|.$$

Note that this dominating function is independent of $\nu \in \mathbb{N}$, but depends on $\ell \in \mathbb{N}$. That is why we have to apply a generalized version of Lebesgue's dominated convergence theorem, that is Theorem C.1, which nevertheless yields

$$\lim_{\ell \to \infty} \|\varphi'_\nu(F_\ell)DF - \varphi'_\nu(F)DF\|_{L^p(\Omega;L^2([0,T];L_2(U_0,\tilde{H})))}$$
$$= \lim_{\ell \to \infty} \left(\mathbf{E}\left[\left(\int_0^T \sum_{\mu=1}^\infty \|(\varphi'_\nu(F_\ell) - \varphi'_\nu(F))DF(t)e_\mu\|_{\tilde{H}}^2 \, dt\right)^{\frac{p}{2}}\right]\right)^{\frac{1}{p}} = 0.$$

For the third summand we use the fact that

$$\lim_{\nu \to \infty} \|(\varphi'_\nu(h_1) - \varphi'(h_1))h_2\|_{\tilde{H}}^2 = \lim_{\nu \to \infty} \sum_{\eta=\nu+1}^\infty |(\varphi'(h_1)h_2, \tilde{e}_\eta)_{\tilde{H}}|^2 = 0$$

for all $h_1, h_2 \in H$. From this we obtain

$$\lim_{\nu \to \infty} \|(\varphi'_\nu(F(\omega)) - \varphi'(F(\omega)))DF(t,\omega)e_\mu\|_{\tilde{H}} = 0$$

for almost all $\omega \in \Omega$, $t \in [0,T]$ and all $\mu \in \mathbb{N}$. From (4.15) we obtain the dominating function

$$\|(\varphi'_\nu(F(\omega)) - \varphi'(F(\omega)))DF(t,\omega)e_\mu\|_{\tilde{H}} \leq C\left(1 + 2\|F(\omega)\|^q\right)\|DF(t,\omega)e_\mu\|$$

and we conclude

$$\lim_{\nu \to \infty} \|\varphi'_\nu(F)DF - \varphi'(F)DF\|_{L^p(\Omega;L^2([0,T];L_2(U_0,\tilde{H})))} = 0$$

by Lebesgue's dominated convergence theorem. This completes the proof. $\square$

For a stochastic process $f : [0,T] \times \Omega \to H$ such that $f \in \mathbb{D}^{1,p}(L^2([0,T];H))$ we recall that Df is a stochastic process which maps into the space $L_2(U_0, L^2([0,T];H))$. Consequently, by (4.8) it follows that $Df(t;u)$ is an H-valued stochastic process and we set

$$Df(\sigma, t; u) := [Df(t;u)](\sigma) \quad \text{for } \sigma, t \in [0,T], \ u \in U_0. \tag{4.16}$$

A different way of understanding this notation is by first evaluating the stochastic process at time σ and then taking the Malliavin derivative. That is, $f(\sigma) \in \mathbb{D}^{1,p}(H)$ for almost all $\sigma \in [0,T]$ and

$$[Df(\sigma)](t;u) = Df(\sigma, t; u) \quad \text{for } t \in [0,T], \ u \in U_0.$$

As demonstrated in the proof of Proposition 4.8 the last equality is proved in the usual way by first approximating f by smooth and cylindrical random variables in $\mathscr{S}(L^2([0, T]; H))$ and then by showing that the Malliavin derivative and the evaluation at a given time point $\sigma \in [0, T]$ commute on $\mathscr{S}(L^2([0, T]; H))$.

The next result is concerned with the interplay between the derivative operator and the Bochner integral.

Proposition 4.8. *Consider a stochastic process* $f : [0, T] \times \Omega \to H$ *such that* $f \in \mathbb{D}^{1,p}(L^2([0, T]; H))$ *with* $p \geq 2$. *Then we have*

$$\int_0^T f(\sigma) \, d\sigma \in \mathbb{D}^{1,p}(H),$$

and

$$\left[D \int_0^T f(\sigma) \, d\sigma \right] (t; u) = \int_0^T Df(\sigma, t; u) \, d\sigma \qquad (4.17)$$

for all $u \in U_0$ *and almost all* $t \in [0, T]$.

Proof. First, we assume that $f \in \mathscr{S}(L^2([0, T]; H))$. In this case, there exists a representation of the form (4.1), where now $(h_j)_{j=1,\dots,n} \subset L^2([0, T]; H)$. Then, let us introduce the random variable $F \in L^p(\Omega; H)$ which is given by

$$F := \int_0^T f(\sigma) \, d\sigma = \sum_{j=1}^n f_j(W(\Phi_1), \dots, W(\Phi_m)) \int_0^T h_j(\sigma) \, d\sigma.$$

Hence, $F \in \mathscr{S}(H)$ and, for all $u \in U_0$ and almost all $t \in [0, T]$,

$$DF(t; u) = \sum_{j=1}^n \sum_{i=1}^m \frac{\partial}{\partial x_i} f_j(W(\Phi_1), \dots, W(\Phi_m)) \int_0^T h_j(\sigma) \, d\sigma \otimes \Phi_i(t) u$$

$$= \int_0^T \sum_{j=1}^n \sum_{i=1}^m \frac{\partial}{\partial x_i} f_j(W(\Phi_1), \dots, W(\Phi_m)) h_j(\sigma) \Phi_i(t) u \, d\sigma$$

$$= \int_0^T Df(\sigma, t; u) \, d\sigma.$$

Therefore, the assertion is true for all $f \in \mathscr{S}(L^2([0, T]; H))$.

For $f \in \mathbb{D}^{1,p}(L^2([0, T]; H))$ we take a sequence $(f^n)_{n \in \mathbb{N}} \subset \mathscr{S}(L^2([0, T]; H))$ such that $f^n \to f$ as $n \to \infty$ with respect to the norm in $\mathbb{D}^{1,p}(L^2([0, T]; H))$. Then, it holds that

$$\left\| \int_0^T f^n(\sigma)\,d\sigma - \int_0^T f(\sigma)\,d\sigma \right\|_{L^p(\Omega;H)} \le \left(\mathbf{E}\left[\left(\int_0^T \|f^n(\sigma) - f(\sigma)\|^2\,d\sigma \right)^{\frac{p}{2}} \right] \right)^{\frac{1}{p}}$$

$$\le C\|f^n - f\|_{L^p(\Omega;L^2([0,T];H))}$$

$$\le C\|f^n - f\|_{\mathbb{D}^{1,p}(L^2([0,T];H))}.$$

Hence, since D is a closed operator, it remains to show that

$$\left[D\int_0^T f^n(\sigma)\,d\sigma \right](t;u) \to \int_0^T Df(\sigma,t;u)\,d\sigma \qquad \text{as } n \to \infty$$

for all $u \in U_0$ and almost all $t \in [0,T]$. Indeed, we have

$$\left\| D\int_0^T f^n(\sigma)\,d\sigma - \int_0^T Df(\sigma,\cdot;\cdot)\,d\sigma \right\|_{L^p(\Omega;L^2([0,T];L_2^0))}$$

$$\le C\left(\mathbf{E}\left[\left(\int_0^T \|D(f^n(\sigma) - f(\sigma))\|^2_{L^2([0,T];L_2^0)}\,d\sigma \right)^{\frac{p}{2}} \right] \right)^{\frac{1}{p}}$$

$$= C\|Df^n - Df\|_{L^p(\Omega;L^2([0,T];L_2(U_0,L^2([0,T];H))))}$$

$$\le C\|f^n - f\|_{\mathbb{D}^{1,p}(L^2([0,T];H))}.$$

Since $f^n \to f$ in $\mathbb{D}^{1,p}(L^2([0,T];H))$ the proof is complete. $\qquad\qquad \square$

4.2 The Divergence Operator

The main object of this section is to establish the integration by parts formula in Proposition 4.3 for a much larger class of stochastic integrands Ψ. For this we introduce the divergence operator δ, which we define in the same way as in [32, Def. 2.7] and [54]:

Definition 4.9. The adjoint $\delta\colon \mathrm{dom}(\delta) \subset L^2(\Omega;L^2([0,T];L_2^0)) \to L^2(\Omega;H)$ of the Malliavin derivative is called *divergence operator*. In particular, its domain $\mathrm{dom}(\delta)$ consists of all $\Psi \in L^2(\Omega;L^2([0,T];L_2^0))$ such that there exists a constant $C = C(\Psi) > 0$ with

$$\left| \mathbf{E}\left[(DF,\Psi)_{L^2([0,T];L_2^0)} \right] \right| \le C\|F\|_{L^2(\Omega;H)} \quad \text{for all } F \in \mathbb{D}^{1,2}(H).$$

For all such $\Psi \in \mathrm{dom}(\delta)$, $\delta(\Psi)$ is defined by the representation theorem as the unique element in $L^2(\Omega;H)$ which satisfies

$$\mathbf{E}\left[(DF,\Psi)_{L^2([0,T];L_2^0)} \right] = \mathbf{E}\left[(F,\delta(\Psi)) \right] \quad \text{for all } F \in \mathbb{D}^{1,2}(H). \qquad (4.18)$$

The divergence operator is also called H-valued *Skorohod* integral and in Proposition 4.12 we will show that δ is an extension of the stochastic Itô-integral. But before we do this, we first record the following properties of δ, which are also well-known in the real-valued situation [58, Sect. 1.3.1].

Proposition 4.10. *The operator* $\delta: \text{dom}(\delta) \subset L^2(\Omega; L^2([0, T]; L_2^0)) \to L^2(\Omega; H)$ *enjoys the following properties:*

(i) for all $\Psi \in \text{dom}(\delta)$ *it holds*

$$\mathbf{E}[\delta(\Psi)] = 0,$$

(ii) δ *is a closed linear operator on* $\text{dom}(\delta)$.

Proof. (i) Let $\Psi \in \text{dom}(\delta)$. For arbitrary $h \in H$ we consider the smooth and cylindrical random variable $F \equiv h \in \mathscr{S}(H)$. Then $DF = 0$ and, consequently,

$$0 = \mathbf{E}\big[(DF, \Psi)_{L^2([0,T];L_2^0)}\big] = \mathbf{E}\big[(F, \delta(\Psi))\big] = \big(h, \mathbf{E}[\delta(\Psi)]\big).$$

Since $h \in H$ is arbitrary, this proves $\mathbf{E}[\delta(\Psi)] = 0$.

(ii) We show that δ is closed: Consider a sequence $(\Psi_n)_{n \in \mathbb{N}} \subset \text{dom}(\delta)$ such that $\Psi_n \to \Psi \in L^2(\Omega; L^2([0, T]; L_2^0))$. Further, assume that $(\delta(\Psi_n))_{n \in \mathbb{N}}$ converges to $G \in L^2(\Omega; H)$. Then we get

$$\mathbf{E}\big[(DF, \Psi)_{L^2([0,T];L_2^0)}\big] = \lim_{n \to \infty} \mathbf{E}\big[(DF, \Psi_n)_{L^2([0,T];L_2^0)}\big]$$

$$= \lim_{n \to \infty} \mathbf{E}\big[(F, \delta(\Psi_n))\big] = \mathbf{E}\big[(F, G)\big]$$

for all $F \in \mathbb{D}^{1,2}(H)$.

From this we obtain $\Psi \in \text{dom}(\delta)$ and $\delta(\Psi) = G$. $\square$

The next lemma is a slightly modified version of [32, Lem. 2.9]. For its formulation it is convenient to introduce the following notation: For $F \in \mathscr{S}(H)$ of the form (4.1) and $\Phi \in L^2([0, T]; L_2(U_0, \mathbb{R}))$ we set

$$D_\Phi F := \sum_{j=1}^{n} \sum_{i=1}^{m} (\Phi_i, \Phi)_{L^2([0,T];L_2(U_0;\mathbb{R}))} \frac{\partial}{\partial x_i} f_j(W(\Phi_1), \dots, W(\Phi_m)) h_j.$$

$$(4.19)$$

Lemma 4.11. *Let* $G \in \mathscr{S}(H)$ *and* $\Phi \in L^2([0, T]; L_2(U_0, \mathbb{R}))$. *Then, it holds that* $G \otimes \Phi \in \text{dom}(\delta)$ *and* $\delta(G \otimes \Phi) = W(\Phi)G - D_\Phi G$.

Proof. The proof follows [32, Lem. 2.9]. For $G \in \mathscr{S}(H)$ and an arbitrary $F \in \mathscr{S}(H)$ we choose a representation of the form (4.1) with respect to the same orthonormal systems $(h_j)_{j=1,...,n} \subset H$ and $(\Phi_i)_{i=1,...,m} \subset L^2([0,T]; L_2(U_0, \mathbb{R}))$. Then, we have

$$
\mathbf{E}\big[\big(DF, G \otimes \Phi\big)_{L^2([0,T];L_2^0)}\big]
$$

$$
= \sum_{j,\ell=1}^{n} \sum_{i=1}^{m} (h_j \otimes \Phi_i, h_\ell \otimes \Phi)_{L^2([0,T];L_2^0)}
$$

$$
\times \mathbf{E}\Big[\Big(\frac{\partial}{\partial x_i} f_j(W(\Phi_1), \ldots, W(\Phi_m))\Big) g_\ell(W(\Phi_1), \ldots, W(\Phi_m))\Big]
$$

$$
= \sum_{j=1}^{n} \sum_{i=1}^{m} (\Phi_i, \Phi)_{L^2([0,T];L_2(U_0,\mathbb{R}))}
$$

$$
\times \Big(\mathbf{E}\Big[\frac{\partial}{\partial x_i}\big(f_j(W(\Phi_1), \ldots, W(\Phi_m)) g_j(W(\Phi_1), \ldots, W(\Phi_m))\big)\Big]
$$

$$
- \mathbf{E}\Big[f_j(W(\Phi_1), \ldots, W(\Phi_m)) \frac{\partial}{\partial x_i} g_j(W(\Phi_1), \ldots, W(\Phi_m))\Big]\Big)
$$

$$
= \sum_{j=1}^{n} \mathbf{E}\big[f_j(W(\Phi_1), \ldots, W(\Phi_m)) g_j(W(\Phi_1), \ldots, W(\Phi_m)) W(\Phi)\big]
$$

$$
- \mathbf{E}\big[(F, D_\Phi G)\big]
$$

$$
= \mathbf{E}\big[(F, W(\Phi)G - D_\Phi G)\big],
$$

where we applied (4.19) and Proposition 4.3 with $H = \mathbb{R}$ in the third step.

Since this is true for all $F \in \mathscr{S}(H)$, it follows that $G \otimes \Phi \in \text{dom}(\delta)$ and $\delta(G \otimes \Phi) = W(\Phi)G - D_\Phi G$. $\qquad\square$

The next proposition is found in [32, Lem. 2.10]. In order to keep this section self-contained we also cite its proof.

Proposition 4.12. *Let* $\Psi \in L^2(\Omega; L^2([0,T]; L_2^0))$ *be a predictable stochastic process with values in* L_2^0. *Then* $\Psi \in \text{dom}(\delta)$ *and*

$$
\delta(\Psi) = \int_0^T \Psi(t) \, dW(t),
$$

that is, $\delta(\Psi) \in L^2(\Omega; H)$ *coincides with the stochastic Itô-integral.*

Proof. We first consider a stochastic process $\Psi \in L^2(\Omega; L^2([0,T]; L_2^0))$ of the form $\Psi = G \otimes \Phi$ with $G \in \mathscr{S}(H)$ and $\Phi \in L^2([0,T]; L_2(U_0, \mathbb{R}))$. Additionally, we assume that Φ is left-continuous and that there exists an $s \in [0,T]$ such that G is $\mathscr{F}_s$-measurable and $\text{supp}(\Phi) \subset (s, T]$. Under these conditions Ψ is predictable. Therefore, the Itô-integral of Ψ exists and is given by

$$\int_0^T \Psi(t)\,dW(t) = \int_s^T \Phi(t)\,dW(t)\,G = W(\Phi)G.$$

On the other hand, Lemma 4.11 yields

$$\delta(\Psi) = W(\Phi)G - D_\Phi G.$$

Now, let G have the representation

$$G = \sum_{j=1}^n g_j(W(\Phi_1), \dots, W(\Phi_m))h_j.$$

Since G is $\mathscr{F}_s$-measurable, we may assume that $\mathrm{supp}(\Phi_i) \subset [0, s]$ for all $i = 1, \dots, m$. Hence, for $i = 1, \dots, m$,

$$\left(\Phi_i, \Phi\right)_{L^2([0,T];L_2(U_0,\mathbb{R}))} = 0$$

and, therefore, $D_\Phi G = 0$. This proves that the divergence operator and the Itô-integral coincide for this class of processes.

By $\mathscr{E}$ we denote the linear span of all simple predictable stochastic processes of the form described above. Note that $\mathscr{E}$ is dense in the space of all predictable and square-integrable stochastic processes which take values in L_2^0.

From this, together with the fact that δ is a closed linear operator (Proposition 4.10(ii)), we obtain the assertion for arbitrary predictable stochastic processes $\Psi \in L^2(\Omega; L^2([0, T]; L_2^0))$.   □

Combining (4.18) and Proposition 4.12 we obtain what is often called *Bismut's integration by parts formula*.

Theorem 4.13. *For all* $\Psi \in N_W^2(0, T; H)$ *and all* $F \in \mathbb{D}^{1,2}(H)$ *it holds that*

$$\mathbf{E}\left[\left(F, \int_0^T \Psi(t)\,dW(t)\right)\right] = \mathbf{E}\left[(DF, \Psi)_{L^2([0,T];L_2^0)}\right]. \tag{4.20}$$

The remainder of this section is concerned with the interplay of the derivative operator and the divergence operator, we recall the notation (4.8).

Lemma 4.14. *Consider* $G_1, G_2 \in \mathscr{S}^2(H)$ *and* $\Psi_1, \Psi_2 \in L^2([0, T]; L_2(U_0, \mathbb{R}))$. *Then it holds that* $\delta(G_i \otimes \Psi_i) \in \mathbb{D}^{1,2}(H)$, *for all* $i = 1, 2$, *and*

$$D(\delta(G_i \otimes \Psi_i))(t; u) = G_i \otimes \Psi_i(t)u + \delta(DG_i(t; u) \otimes \Psi_i), \tag{4.21}$$

for all $u \in U_0$ *and almost all* $t \in [0, T]$. *Further, we have*

$$\mathbf{E}\big[\big(\delta(G_1 \otimes \Psi_1), \delta(G_2 \otimes \Psi_2)\big)\big]$$

$$= \mathbf{E}\big[(G_1 \otimes \Psi_1, G_2 \otimes \Psi_2)_{L^2([0,T];L_2^0)}\big]$$

$$+ \int_{[0,T]^2} \sum_{k_1,k_2=1}^{\infty} \mathbf{E}\big[\big(DG_1(t_2, e_{k_2})\Psi_1(t_1)e_{k_1}, DG_2(t_1, e_{k_1})\Psi_2(t_2)e_{k_2}\big)\big] \, \mathrm{d}t_1 \, \mathrm{d}t_2,$$

$$(4.22)$$

where $(e_k)_{k\in\mathbb{N}}$ denotes an arbitrary orthonormal basis of U_0.

Proof. Under the given assumption on G_i and Ψ_i, $i = 1, 2$, it follows from Lemma 4.11 that $G_i \otimes \Psi_i \in \mathrm{dom}(\delta)$ and

$$\delta(G_i \otimes \Psi_i) = W(\Psi_i)G_i - D_{\Psi_i}G_i.$$

Hence, by the additional smoothness assumption on G_i and (4.19) we get that $\delta(G_i \otimes \Psi_i) \in \mathscr{S}(H) \subset \mathbb{D}^{1,2}(H)$. A further application of Lemma 4.11 yields

$$D(\delta(G_i \otimes \Psi_i))(t;u) = G_i \otimes \Psi_i(t)u + W(\Psi_i)DG_i(t;u) - D_{\Psi_i}(DG_i(t;u))$$

$$= G_i \otimes \Psi_i(t)u + \delta(DG_i(t;u) \otimes \Psi_i),$$

where we used that $DG_i(t;u) \in \mathscr{S}(H)$ and

$$D(D_{\Psi_i}G_i)(t;u) = D_{\Psi_i}(DG_i(t;u)), \quad \text{for all } t \in [0,T], \ u \in U_0.$$

Thus, it remains to prove (4.22). For this we first apply (4.18) and obtain

$$\mathbf{E}\big[\big(\delta(G_1 \otimes \Psi_1), \delta(G_2 \otimes \Psi_2)\big)\big] = \mathbf{E}\big[(G_1 \otimes \Psi_1, D(\delta(G_2 \otimes \Psi_2)))_{L^2([0,T],L_2^0)}\big]$$

$$= \mathbf{E}\big[(G_1 \otimes \Psi_1, G_2 \otimes \Psi_2)_{L^2([0,T],L_2^0)}\big]$$

$$+ \mathbf{E}\big[(G_1 \otimes \Psi_1, \delta(DG_2(\cdot;\cdot) \otimes \Psi_2))_{L^2([0,T],L_2^0)}\big].$$

A further application of (4.18) to the last summand yields

$$\mathbf{E}\big[(G_1 \otimes \Psi_1, \delta(DG_2(\cdot;\cdot) \otimes \Psi_2))_{L^2([0,T],L_2^0)}\big]$$

$$= \int_0^T \sum_{k_1=1}^{\infty} \mathbf{E}\big[(G_1 \otimes \Psi_1(t_1)e_{k_1}, \delta(DG_2(t_1;e_{k_1}) \otimes \Psi_2))\big] \, \mathrm{d}t_1$$

$$= \int_{[0,T]^2} \sum_{k_1,k_2=1}^{\infty} \mathbf{E}\big[(DG_1(t_2;e_{k_2}) \otimes \Psi_1(t_1)e_{k_1}, DG_2(t_1;e_{k_1}) \otimes \Psi_2(t_2)e_{k_2})\big] \mathrm{d}t_1 \, \mathrm{d}t_2.$$

This completes the proof.   □

As the next proposition shows, the domain of δ contains many more stochastic processes, which are not necessarily adapted. This result is shown in [58, Prop. 1.3.1] in the real-valued situation. Compare also with [32, Prop. 3.2]. For the proof it is convenient to introduce the space

$$\mathscr{E}^2 := \text{span}\{G \otimes \Psi : G \in \mathscr{S}^2(H), \ \Psi \in L^2([0,T]; L_2(U_0, \mathbb{R}))\}.$$

It holds that the space $\mathscr{E}^2$ is dense in $\mathbb{D}^{2,2}(L^2([0,T]; L_2^0))$ and, since $\mathscr{S}^2(H)$ is a dense subspace of $\mathscr{S}(H)$, it is also clear that $\mathscr{E}^2$ is dense in $\mathbb{D}^{1,2}(L^2([0,T]; L_2^0))$.

Proposition 4.15. *Let* $\Upsilon \in \mathbb{D}^{1,2}(L^2([0,T]; L_2^0))$. *Then* $\Upsilon \in \text{dom}(\delta)$ *and*

$$\mathbf{E}\big[\|\delta(\Upsilon)\|^2\big] \leq \mathbf{E}\big[\|\Upsilon\|^2_{L^2([0,T];L_2^0)}\big] + \mathbf{E}\big[\|D\Upsilon\|^2_{L^2([0,T]^2;L_2(U_0,L_2^0))}\big]. \tag{4.23}$$

Proof. For $\Upsilon \in \mathbb{D}^{1,2}(L^2([0,T]; L_2^0))$ take a sequence $(\Upsilon_n)_{n \in \mathbb{N}} \subset \mathscr{E}^2$ such that $\Upsilon_n \to \Upsilon$ in $\mathbb{D}^{1,2}(L^2([0,T]; L_2^0))$ as $n \to \infty$. By Lemma 4.11 it follows that $\Upsilon_n \in \text{dom}(\delta)$ for every $n \in \mathbb{N}$ and (4.22) yields (4.23) for all $\Upsilon_n \in \mathscr{E}^2$.

If we apply (4.23) to $\Upsilon_n - \Upsilon_m$, we get that $(\delta(\Upsilon_n))_{n \in \mathbb{N}}$ is a Cauchy sequence in $L^2(\Omega; H)$. Since δ is a closed operator, it follows that $\Upsilon \in \text{dom}(\delta)$ and $\delta(\Upsilon_n) \to \delta(\Upsilon)$ as $n \to \infty$ and (4.23) remains valid. $\square$

Proposition 4.16. *For every* $\Upsilon \in \mathbb{D}^{2,2}(L^2([0,T]; L_2^0))$ *it holds that* $\delta(\Upsilon) \in \mathbb{D}^{1,2}(H)$ *and*

$$[D\delta(\Upsilon)](t;u) = \Upsilon(t)u + \delta(D\Upsilon(t;u)) \tag{4.24}$$

for all $u \in U_0$ *and almost all* $t \in [0,T]$.

Proof. First, we note that $D\Upsilon(t;u) \in \mathbb{D}^{1,2}(L^2([0,T]; L_2^0))$ for all $u \in U_0$ and almost all $t \in [0,T]$. Therefore, by Proposition 4.15 it holds that $D\Upsilon(t;u) \in \text{dom}(\delta)$ and the right hand side in (4.24) is well-defined.

As in the proof of Proposition 4.15 we take a sequence $(\Upsilon_n)_{n \in \mathbb{N}} \subset \mathscr{E}^2$ such that $\Upsilon_n \to \Upsilon$ in $\mathbb{D}^{2,2}(L^2([0,T]; L_2^0))$ as $n \to \infty$. Then it holds that $\delta(\Upsilon_n) \to \delta(\Upsilon)$ in $L^2(\Omega; H)$. In addition for all $\Upsilon_n \in \mathscr{E}^2$ we obtain from Lemma 4.14 that $\delta(\Upsilon_n) \in \mathbb{D}^{1,2}(H)$ and (4.24) is valid by (4.21).

An application of (4.23) yields

$$\big\|\Upsilon + \delta(D\Upsilon(\cdot;\cdot)) - D(\delta(\Upsilon_n))\big\|_{L^2(\Omega;L^2([0,T];L_2^0))}$$

$$\leq \big\|\Upsilon - \Upsilon_n\big\|_{L^2(\Omega;L^2([0,T];L_2^0))} + \Big(\int_0^T \sum_{k=1}^\infty \mathbf{E}\big[\|\delta(D[(\Upsilon - \Upsilon_n)](t;e_k))\|^2\big]\,dt\Big)^{\frac{1}{2}}$$

$$\leq \big\|\Upsilon - \Upsilon_n\big\|_{L^2(\Omega;L^2([0,T];L_2^0))}$$

$$+ \left(\int_0^T \sum_{k=1}^{\infty} \mathbf{E}\big[\big\| [D(\Upsilon - \Upsilon_n)](t;e_k) \big\|_{L^2([0,T];L_2^0)}^2 \big] \, dt \right)^{\frac{1}{2}}$$

$$+ \left(\int_0^T \sum_{k=1}^{\infty} \mathbf{E}\big[\big\| D\big([D(\Upsilon - \Upsilon_n)](t;e_k) \big) \big\|_{L^2([0,T]^2;L_2(U_0;L_2^0))}^2 \big] \, dt \right)^{\frac{1}{2}}$$

$$\leq 3 \| \Upsilon - \Upsilon_n \|_{\mathbb{D}^{2,2}(L^2([0,T];L_2^0))},$$

where $(e_k)_{k \in \mathbb{N}}$ denotes an arbitrary orthonormal basis in U_0. Hence, by our choice of the sequence $(\Upsilon_n)_{n \in \mathbb{N}}$ it follows that

$$[D\delta(\Upsilon_n)])(t;u) \to \Upsilon + \delta(D\Upsilon(t;u)) \quad \text{as } n \to \infty$$

for all $u \in U_0$ and almost all $t \in [0, T]$. Since the derivative operator is closed, we get $\delta(\Upsilon) \in \mathbb{D}^{1,2}(H)$ and (4.24) is valid. □

Remark 4.17. Let us remark that from (4.23) it follows that the divergence operator δ is continuous from $\mathbb{D}^{1,2}(L^2([0,T]; L_2^0))$ to $L^2(\Omega; H)$. Additionally, the arguments given in the proof of Proposition 4.16 are sufficient to show that δ is also continuous as a linear operator from $\mathbb{D}^{2,2}(L^2([0,T]; L_2^0))$ to $\mathbb{D}^{1,2}(H)$. For a more general result in this direction we refer to [58, Prop. 1.5.7].

We also note that [58, Prop. 1.3.2] gives a similar result as Proposition 4.16 but under slightly weaker smoothness assumptions on Ψ.

4.3 A Short Proof of the Stochastic Fubini Theorem

As a first application of Bismut's integration by parts formula (4.20) we present a short proof of a stochastic Fubini theorem. For this let us recall from Sect. 2.2 that by $\mathscr{P}_T$ we denote the σ-field of all predictable stochastic processes.

In [18, Th. 4.18] one can find a stochastic Fubini theorem, whose proof takes more than three pages and uses approximations of a predictable process by a suitable class of elementary processes. By using the integration by parts formula (4.20) the same result is established in a shorter and simplified proof.

We also refer to [68] for a more recent presentation of the stochastic Fubini theorem, which also contains some generalizations. However, the proof is based on similar techniques as in [18, Th. 4.18].

Theorem 4.18. *Let $(E, \mathscr{E}, \mu)$ be a measure space with finite measure μ and let $\Phi: [0, T] \times \Omega \times E \to L_2^0$ be a $(\mathscr{P}_T \times \mathscr{E})/\mathscr{B}(L_2^0)$-measurable mapping. Under the condition*

$$\int_E \left(\mathbf{E}\Big[\int_0^T \| \Phi(t,x) \|_{L_2^0}^2 \, dt \Big] \right)^{\frac{1}{2}} \, d\mu(x) < \infty, \tag{4.25}$$

it holds that

$$\int_0^T \int_E \Phi(t,x)\,d\mu(x)\,dW(t) = \int_E \int_0^T \Phi(t,x)\,dW(t)\,d\mu(x) \quad \mathbf{P}\text{-}a.s. \quad (4.26)$$

In particular, all integrals in (4.26) are well-defined.

Proof. As in [18, Th. 4.18] let us first note that by the classical Fubini theorem the integral

$$\int_E \Phi(\cdot,x)\,d\mu(x)$$

is $\mathscr{P}_T / \mathscr{B}(L_2^0)$-measurable and it holds by (4.25) that

$$\left(\mathbf{E}\left[\int_0^T \left\| \int_E \Phi(t,x)\,d\mu(x) \right\|_{L_2^0}^2 dt \right] \right)^{\frac{1}{2}}$$

$$\leq \int_E \left(\mathbf{E}\left[\int_0^T \|\Phi(t,x)\|_{L_2^0}^2 dt \right] \right)^{\frac{1}{2}} d\mu(x) < \infty.$$

Consequently, the iterated integral on the left hand side in (4.26) is well-defined.

Below we will convince ourselves that all transformations in the following chain of equalities are rigorous. For arbitrary $F \in \mathbb{D}^{1,2}(H)$ it holds that

$$\mathbf{E}\left[\left(F, \int_0^T \int_E \Phi(t,x)\,d\mu(x)\,dW(t) \right) \right]$$

$$= \mathbf{E}\left[\int_0^T \left(DF(t), \int_E \Phi(t,x)\,d\mu(x) \right)_{L_2^0} dt \right] \quad (4.27)$$

$$= \int_E \mathbf{E}\left[\int_0^T \left(DF(t), \Phi(x,t) \right)_{L_2^0} dt \right] d\mu(x) \quad (4.28)$$

$$= \int_E \mathbf{E}\left[\left(F, \int_0^T \Phi(x,t)\,dW(t) \right) \right] d\mu(x) \quad (4.29)$$

$$= \mathbf{E}\left[\left(F, \int_E \int_0^T \Phi(x,t)\,dW(t)\,d\mu(x) \right) \right]. \quad (4.30)$$

Since $\mathbb{D}^{1,2}(H)$ is dense in H, this proves (4.26) provided that the iterated integral in (4.30) is well-defined. We will come back to this point at the end of the proof.

In (4.27) we applied Theorem 4.13. For (4.28) we define the linear functional $\ell_1 \colon L^2(\Omega; L^2([0,T]; L_2^0)) \to \mathbb{R}$ by

$$\ell_1(\Psi) = \mathbf{E}\left[\int_0^T \left(DF(t), \Psi(t) \right)_{L_2^0} dt \right].$$

Since $\|\ell_1\| \leq \|DF\|_{L^2(\Omega;L^2([0,T];L_2^0))} < \infty$ we can interchange the Bochner integral
in (4.27) with ℓ_1 (cf. [14, Prop. E.11]). In particular, we stress that the mapping
$x \mapsto \ell_1(\Phi(\cdot, x))$ is $\mathscr{E}/\mathscr{B}(\mathbb{R})$-measurable for every choice of $F \in \mathbb{D}^{1,2}(H)$ (see [14,
Theorem E.9]).

For (4.29) we first note that under the given assumptions on Φ the x-section
$\Phi(\cdot, x)$ is a square-integrable predictable stochastic process for μ-almost all $x \in E$.
Thus, the equality (4.29) is another application of Theorem 4.13.

Next, we introduce the linear functional $\ell_2 : L^2(\Omega; H) \to \mathbb{R}$ which is given by

$$\ell_2(\Upsilon) = \mathbf{E}[(F, \Upsilon)].$$

Note that the mapping $x \mapsto \ell_2(\int_0^T \Phi(x, t) \, dW(t))$ is for μ-almost every $x \in E$
equal to the mapping $x \mapsto \ell_1(\Phi(\cdot, x))$ and therefore $\mathscr{E}/\mathscr{B}(\mathbb{R})$-measurable.

Since this is true for all linear functionals ℓ_1 and ℓ_2 which are generated by
arbitrary $F \in \mathbb{D}^{1,2}(H)$, we obtain by Cohn [14, Th. E.9] that also the mapping
$x \mapsto \int_0^T \Phi(x, t) \, dW(t)$ is $\mathscr{E}/\mathscr{B}(L^2(\Omega; H))$-measurable and, by (4.25), Bochner
integrable. Therefore, the iterated integral on the right hand side of (4.26) is well-
defined and the equality (4.30) follows from interchanging the linear functional ℓ_2
with the Bochner integral.   □

Example 4.19 (Continuation of 2.24). In Example 2.24 we embedded a simple
SODE in our framework of semilinear stochastic evolution equations. As a result
we obtained the representation (2.25) of a real-valued Wiener process

$$W(t) = \lambda \int_0^t e^{-\lambda(t-\sigma)} W(\sigma) \, d\sigma + \int_0^t e^{-\lambda(t-\sigma)} \, dW(\sigma) \quad \text{for all } t \in [0, T].$$

$$(4.31)$$

In this example we prove that this formula also holds for more general Hilbert
space-valued Q-Wiener processes on $H = U$ and arbitrary strongly continuous
semigroups $(E(t))_{t \in [0,T]}$ on H with generator $-A$. Here, we assume that $Q \in L(H)$
is of finite trace and hence $\|E(t)\|_{L_2^0}^2 \leq \|E(t)\|_{L(H)}^2 \mathrm{Tr}(Q)$.

Fix $t \in [0, T]$ and consider the stochastic integral

$$\int_0^t \left(E(t - \sigma) - \mathrm{Id}_H\right) dW(\sigma) = \int_0^t (-A) \int_0^{t-\sigma} E(\tau) \, d\tau \, dW(\sigma)$$

$$= -A \int_0^t \int_0^t 1_{[0,t-\sigma]}(\tau) E(\tau) \, d\tau \, dW(\sigma),$$

where we applied Lemma B.5(ii). Since the mapping $\Phi : [0, t] \times [0, t] \to L_2^0$,
$(\sigma, \tau) \mapsto 1_{[0,t-\sigma]}(\tau) E(\tau)$ is deterministic and left-continuous for fixed τ, it is
predictable. Therefore, the stochastic Fubini theorem is applicable and yields

$$\int_0^t \left(E(t-\sigma) - \mathrm{Id}_H \right) \mathrm{d}W(\sigma) = -A \int_0^t \int_0^t 1_{[0,t-\sigma]}(\tau) E(\tau)\, \mathrm{d}W(\sigma)\, \mathrm{d}\tau$$

$$= -A \int_0^t E(\tau) \int_0^{t-\tau} \mathrm{d}W(\sigma)\, \mathrm{d}\tau$$

$$= -A \int_0^t E(t-\tau) W(\tau)\, \mathrm{d}\tau,$$

or equivalently,

$$W(t) = A \int_0^t E(t-\sigma) W(\sigma)\, \mathrm{d}\sigma + \int_0^t E(t-\sigma)\, \mathrm{d}W(\sigma),$$

which is the infinite dimensional analogue of (4.31).

Chapter 5
A Malliavin Calculus Approach to Weak Convergence

The aim of this chapter is to present a new technique to analyze the weak error of convergence of spatially semidiscrete approximations and spatio-temporal discretizations of the solution of a linear stochastic evolution equation with additive noise.

5.1 Preliminaries

In this chapter we are interested in the mild solution $X : [0, T] \times \Omega \to H$ to the stochastic evolution equation

$$dX(t) + [AX(t) + f(t)]\,dt = g(t)\,dW(t), \quad \text{for } 0 \le t \le T,$$
$$X(0) = x_0 \in H. \tag{5.1}$$

We work under the same conditions as in Sect. 2.3 , but in contrast to Chap. 2 we assume that f and g do not depend on X. We refer to Sect. 5.2 for a precise formulation of all assumptions used throughout this chapter.

Note that (5.1) is called a *linear stochastic evolution equation* with additive noise. Equations of this type are well-studied in the literature [18, Chap. 5] and their mild solutions are given by the variation of constants formula

$$X(t) = E(t)x_0 - \int_0^t E(t - \sigma)f(\sigma)\,d\sigma + \int_0^t E(t - \sigma)g(\sigma)\,dW(\sigma) \quad \text{P–a.s.} \tag{5.2}$$

for all $0 \le t \le T$ provided all integrals are well defined.

Note that (5.2) also is a *weak solution* to (5.1), that is, the stochastic process $X : [0, T] \times \Omega \to H$ satisfies

R. Kruse, *Strong and Weak Approximation of Semilinear Stochastic Evolution Equations*, 109
Lecture Notes in Mathematics 2093, DOI 10.1007/978-3-319-02231-4_5,
© Springer International Publishing Switzerland 2014

$$\left(X(t), \zeta\right) = \left(x_0, \zeta\right) + \int_0^t \left(X(\sigma), A^*\zeta\right) + \left(f(\sigma), \zeta\right) d\sigma$$

$$+ \int_0^t \left(\zeta, g(\sigma) dW(\sigma)\right) \quad \mathbf{P}\text{–a.s.}$$

(5.3)

for every $t \in [0, T]$ and $\zeta \in D(A^*)$, where $(A^*, D(A^*))$ denotes the adjoint of $(A, D(A))$ on H. However, we will not make use of this property and refer for more details to [18, Chap. 5.1] and [61, Def. F.0.3].

For a given mapping $\varphi \in C_p^2(H; \mathbb{R})$ our aim is to compute a good estimate of the real number

$$\mathbf{E}\big[\varphi(X(T))\big].$$

In order to achieve this it is necessary to replace the mild solution X by finite-dimensional approximations.

For this we first consider a family of finite dimensional subspaces $(S_h)_{h \in (0,1]} \subset \dot{H}^1$ as in Chap. 3. We first analyze the *weak error of the spatial semidiscretization* which is given by

$$e_h(T) := \big|\mathbf{E}\big[\varphi(X_h(T)) - \varphi(X(T))\big]\big|, \tag{5.4}$$

where the process $X_h \colon [0, T] \times \Omega \to S_h$, $h \in (0, 1]$, satisfies the stochastic evolution equation

$$dX_h(t) + [A_h X_h(t) + P_h f(t)] dt = P_h g(t) dW(t), \quad \text{for } 0 \le t \le T,$$

$$X_h(0) = P_h x_0. \tag{5.5}$$

Here, A_h is defined in (3.11) and P_h denotes the orthogonal projector onto S_h. The semidiscrete analogue to (5.2) reads now

$$X_h(t) = E_h(t) P_h x_0 - \int_0^t E_h(t - \sigma) P_h f(\sigma) d\sigma + \int_0^t E_h(t - \sigma) P_h g(\sigma) dW(\sigma). \tag{5.6}$$

For linear stochastic evolution equations with deterministic inhomogeneities f and g the same error is analyzed in [45].

We also consider the spatio-temporal discretization of X from Sect. 3.6 which is based on a linearly implicit Euler–Maruyama scheme. By $k \in (0, 1]$ let us denote a fixed time step and by $t_j := jk$, $j = 1, \ldots, N_k$, a temporal grid with $N_k \le T < (N_k + 1)k$, $N_k \in \mathbb{N}$. Then, in the situation of this chapter the scheme is given by the recursion

$$X_h^j - X_h^{j-1} + k\big(A_h X_h^j + P_h f(t_{j-1})\big) = P_h g(t_{j-1}) \Delta W^j, \quad \text{for } j = 1, \ldots, N_k,$$

$$X_h^0 = P_h x_0, \tag{5.7}$$

with the $\mathscr{F}_{t_j}$-measurable Wiener increments $\Delta W^j := W(t_j) - W(t_{j-1})$. From this it follows that X_h^j is an $\mathscr{F}_{t_j}$-adapted random variable.

The *weak error of the spatio-temporal discretization* is defined as

$$e_{kh}(T) := \left| \mathbf{E}\big[\varphi(X_h^{N_k}) - \varphi(X(t_{N_k}))\big]\right|. \tag{5.8}$$

Let us remark, that in the context of weak approximations of SEEqs the implicit Euler–Maruyama scheme is also analyzed, for example, in [21, 22, 46].

5.2 Assumptions

In this section we precisely formulate our assumptions which are used throughout this chapter. We also refer to Sect. 5.5, where we indicate some generalizations.

Let $(\Omega, \mathscr{F}, \mathbf{P})$ be a complete probability space with a normal filtration $\mathscr{F}_t \subset \mathscr{F}$ for all $t \in [0, T]$. As in Sect. 2.1 we denote by W an adapted cylindrical Q-Wiener process, where the covariance operator $Q: U \to U$ is linear, bounded, self-adjoint and positive semidefinite. In order to make the results from Chap. 4 available, we further assume that $\mathscr{F}$ and the filtration $(\mathscr{F}_t)_{t \in [0,T]}$ are generated by the Wiener process W.

Assumption 5.1. The linear operator $A: \mathrm{dom}(A) \subset H \to H$ is densely defined, self-adjoint and positive definite with compact inverse.

As in earlier chapters and in Appendix B.2, the operator $-A$ is the generator of an analytic semigroup $(E(t))_{t \in [0,\infty)}$ of contractions on H.

For the formulation of the next two assumptions let $p \in [2, \infty)$.

Assumption 5.2. The mapping $f: [0, T] \times \Omega \to H$ is a predictable stochastic process with Bochner-integrable trajectories. It holds that $\sup_{t \in [0,T]} \|f(t)\|_{L^p(\Omega;H)} < \infty$ and for some $\delta \in [\frac{1}{2}, 1]$ there exists a constant C such that

$$\left\| f(t_1) - f(t_2) \right\|_{L^p(\Omega;H)} \leq C|t_1 - t_2|^\delta \tag{5.9}$$

for all $t_1, t_2 \in [0, T]$.

Assumption 5.3. The mapping $g: [0, T] \times \Omega \to L_2^0$ is a predictable stochastic process. It holds that $\sup_{t \in [0,T]} \|g(t)\|_{L^p(\Omega;L_2^0)} < \infty$ and for some $\delta \in [\frac{1}{2}, 1]$ there exists a constant C such that

$$\left\| g(t_1) - g(t_2) \right\|_{L^p(\Omega;L_2^0)} \leq C|t_1 - t_2|^\delta \tag{5.10}$$

for all $t_1, t_2 \in [0, T]$.

Assumption 5.4. The initial value x_0 is an arbitrary but deterministic element in H.

Note that under these conditions Assumptions 2.14 and 2.16 are not necessarily fulfilled. However, as indicated in Sect. 2.7, the mappings f and g satisfy the assertion of Lemma 2.26. In particular, all appearing integrals in (5.2) and (5.6) are well-defined and the mild solutions exist.

For the representation of the weak error in Sect. 5.3 we also need that the mild solution X is differentiable in the sense of the Malliavin calculus. To assure this we work under the following additional assumptions on f and g. For the formulation recall the definition of the Malliavin derivative and the notation (4.16).

Assumption 5.5. With the same $p \in [2, \infty)$ as in Assumption 5.2 the stochastic process $f : [0, T] \times \Omega \to H$ satisfies $f \in \mathbb{D}^{1,p}(L^2([0, T]; H))$. Further, we assume that $f(t) \in \mathbb{D}^{1,p}(H)$ for every $t \in [0, T]$ and

$$\sup_{t,\sigma \in [0,T]} \|Df(t, \sigma; \cdot)\|_{L^p(\Omega; L_2^0)} < \infty.$$

Assumption 5.6. With the same $p \in [2, \infty)$ as in Assumption 5.3 the stochastic process $g : [0, T] \times \Omega \to L_2^0$ satisfies $g \in \mathbb{D}^{2,p}(L^2([0, T]; L_2^0))$ and $g(t) \in \mathbb{D}^{1,p}(L_2^0)$ for every $t \in [0, T]$ such that

$$\sup_{t,\sigma \in [0,T]} \|Dg(t, \sigma; \cdot)\|_{L^p(\Omega; L_2^0)} < \infty.$$

In addition, we assume that for all $u \in U_0$ and almost all $t \in [0, T]$ the stochastic process $[0, T] \ni \sigma \mapsto E(T - \sigma)Dg(\sigma, t; u) \in L_2^0$ is predictable.

If f and g are mappings, which do not depend on Ω, then Assumptions 5.5 and 5.6 are always satisfied.

Another example of a stochastic process which satisfies Assumption 5.5 is given in Example 4.6 with $f(t) = W(t)$, since

$$\sup_{t,\sigma \in [0,T]} \|[DW(t)](\sigma)\|_{L^2(\Omega; L_2^0)}^2 = \sup_{t,\sigma \in [0,T]} 1_{[0,t]}(\sigma) \sum_{k=1}^{\infty} \mu_k = \mathrm{Tr}(Q) < \infty.$$

Theorem 5.7. *Under Assumptions 5.1–5.6 it holds that the mild solution X to (5.1) satisfies $X(T) \in \mathbb{D}^{1,p}(H)$ and*

$$[DX(T)](t; u) = -\int_0^T E(T - \sigma)Df(\sigma, t; u)\, d\sigma + E(T - t)g(t)u$$

$$+ \int_0^T E(T - \sigma)Dg(\sigma, t; u)\, dW(\sigma)$$

for all $u \in U_0$ and almost all $t \in [0, T]$.

Proof. The mild solution X is uniquely given by (5.2). Having Assumption 5.4 in mind it suffices to show that

$$\int_0^T E(T-\sigma)f(\sigma)\,d\sigma \in \mathbb{D}^{1,p}(H) \tag{5.11}$$

and

$$\int_0^T E(T-\sigma)g(\sigma)\,dW(\sigma) \in \mathbb{D}^{1,p}(H). \tag{5.12}$$

For (5.11) Lemma 4.7 with $q=0$, (4.16) and Assumption 5.5 yield that the stochastic process $[0,T] \ni \sigma \mapsto E(T-\sigma)f(\sigma)$ is an element in $\mathbb{D}^{1,p}(L^2([0,T];H))$. Thus, Proposition 4.8 gives (5.11) and

$$\left[D\int_0^T E(T-\sigma)f(\sigma)\,d\sigma\right](t,u) = \int_0^T E(T-\sigma)Df(\sigma,t;u)\,d\sigma$$

for all $u \in U_0$ and almost all $t \in [0,T]$.

Similarly, two applications of Lemma 4.7 with $q=0$, (4.16) and Assumption 5.6 yield that the stochastic process $[0,T] \ni \sigma \mapsto E(T-\sigma)g(\sigma)$ is an element in $\mathbb{D}^{2,p}(L^2([0,T];L_2^0)) \subset \mathbb{D}^{2,2}(L^2([0,T];L_2^0))$. Hence, Propositions 4.12 and 4.16 are applicable and we obtain (5.12) with

$$\left[D\int_0^T E(T-\sigma)g(\sigma)\,dW(\sigma)\right](t;u) = E(T-t)g(t)u$$

$$+ \int_0^T E(T-\sigma)Dg(\sigma,t;u)\,dW(\sigma)$$

for all $u \in U_0$ and almost all $t \in [0,T]$. This proves (5.12) with $p=2$. For $p>2$ recall from Lemma B.2 that there exists a constant $M>0$ such that $\|E(t)\| \le M$ for all $t \in [0,T]$. Hence, by also applying Assumption 5.6 and Proposition 2.12 it holds

$$\left\|D\int_0^T E(T-\sigma)g(\sigma)\,dW(\sigma)\right\|_{L^p(\Omega;L^2([0,T];L_2^0))}$$

$$\le \left\|E(T-\cdot)g(\cdot)\right\|_{L^p(\Omega;L^2([0,T];L_2^0))}$$

$$+ \left\|\int_0^T E(T-\sigma)Dg(\sigma,\cdot;\cdot)\,dW(\sigma)\right\|_{L^p(\Omega;L^2([0,T];L_2^0))}$$

$$\le M\|g\|_{L^p(\Omega;L^2([0,T];L_2^0))}$$

$$+ CM\left(\mathbf{E}\left[\left(\int_0^T \|Dg(\sigma,\cdot;\cdot)\|_{L_2(U_0,L^2([0,T];L_2^0))}^2\,d\sigma\right)^{\frac{p}{2}}\right]\right)^{\frac{1}{p}}$$

$$\le CM\|g\|_{\mathbb{D}^{1,p}(L^2([0,T];L_2^0))} < \infty.$$

This completes the proof.   □

A straightforward application of the given arguments also yields the analogous result for the discretization schemes. Here, for the spatio-temporal approximation we apply the same notation as in Sect. 3.6, that is we introduce the mapping $E_{kh} : [0, t_{N_k}) \to L(S_h)$ by

$$E_{kh}(t) := (I + kA_h)^{-j}, \quad \text{if } t \in [t_{j-1}, t_j), \quad j = 1, \dots, N_k. \tag{5.13}$$

Further, let us denote two auxiliary processes by

$$\begin{aligned} f_{kh}(t) &:= P_h f(t_{j-1}), \quad \text{if } t \in (t_{j-1}, t_j], \quad j = 1, \dots, N_k, \\ f_{kh}(0) &:= P_h f(0), \end{aligned} \tag{5.14}$$

and, similarly,

$$\begin{aligned} g_{kh}(t) &:= P_h g(t_{j-1}), \quad \text{if } t \in (t_{j-1}, t_j], \quad j = 1, \dots, N_k, \\ g_{kh}(0) &:= P_h g(0). \end{aligned} \tag{5.15}$$

Then the recursion (5.7) is equivalently written in closed form as

$$\begin{aligned} X_h^j = (I + kA_h)^{-j} P_h x_0 &- \int_0^{t_j} E_{kh}(t_j - \sigma) f_{kh}(\sigma) \, d\sigma \\ &+ \int_0^{t_j} E_{kh}(t_j - \sigma) g_{kh}(\sigma) \, dW(\sigma). \end{aligned} \tag{5.16}$$

Theorem 5.8. *Under Assumptions 5.1–5.6 the semidiscrete approximation (5.5) and the spatio-temporal approximation (5.16) satisfy $X_h(T), X_h^{N_k} \in \mathbb{D}^{1,p}(H)$ with*

$$\big[DX_h(T) \big](t; u) = - \int_0^T E_h(T - \sigma) P_h Df(\sigma, t; u) \, d\sigma + E_h(T - t) P_h g(t) u$$

$$+ \int_0^T E_h(T - \sigma) P_h Dg(\sigma, t; u) \, dW(\sigma)$$

and

$$\big[DX_h^{N_k} \big](t; u) = - \int_0^T E_{kh}(t_j - \sigma) Df_{kh}(\sigma, t; u) \, d\sigma + E_{kh}(T - t) g_{kh}(t) u$$

$$+ \int_0^T E_{kh}(T - \sigma) Dg_{kh}(\sigma, t; u) \, dW(\sigma)$$

for all $u \in U_0$ and almost all $t \in [0, T]$.

5.3   A Useful Representation Formula of the Weak Error

In this section we derive a representation formula of the weak error of convergence which is a generalization of a corresponding result in [45]. In this section we also compare our Malliavin Calculus to the more classical approach via the Kolmogorov equation in [45].

First we are concerned with the weak error (5.4) of the spatial semidiscretization (5.6). As in Sect. 3.3, let us introduce the error operator

$$F_h(t) := E_h(t)P_h - E(t), \quad \text{for } t \in [0, T]. \tag{5.17}$$

We also recall our notation of Fréchet differentiable mappings from Sect. 4.1.

Theorem 5.9. *Let the Assumptions 5.1–5.6 hold with $p \in [2, \infty)$. Given a mapping $\varphi \in C_p^2(H, \mathbb{R})$ such that there exists a constant C with*

$$|\varphi(x)| \leq C\left(1 + \|x\|^p\right), \tag{5.18}$$

$$\|\varphi'(x)\| \leq C\left(1 + \|x\|^{p-1}\right), \tag{5.19}$$

$$\|\varphi''(x)\|_{L(H,H)} \leq C\left(1 + \|x\|^{p-2}\right) \tag{5.20}$$

for all $x \in H$. Then the weak error of the semidiscrete approximation satisfies

$$\mathbf{E}\left[\varphi(X_h(T)) - \varphi(X(T))\right]$$

$$= \int_0^1 \mathbf{E}\left[\left(\varphi'\big(X(T) + s(X_h(T) - X(T))\big), F_h(T)x_0 - \int_0^T F_h(T - \sigma)f(\sigma)\,d\sigma\right)\right]ds$$

$$+ \int_0^1 \int_0^T \mathbf{E}\left[\left(\varphi''\big(X(T) + s(X_h(T) - X(T))\big)\right.\right.$$

$$\left.\left.\times \big((1 - s)[DX(T)](\sigma) + s[DX_h(T)](\sigma), F_h(T - \sigma)g(\sigma)\big)_{L_2^0}\right]d\sigma\,ds.$$

Proof. First, we note that the weak error and all integrals in the representation formula of the weak error are well-defined under (5.18)–(5.20) and in view of Theorems 5.7 and 5.8.

For instance, we have by (5.20) and several applications of the Hölder inequality

$$\int_0^T \mathbf{E}\left[\left|\left(\varphi''\big(X(T) + s(X_h(T) - X(T))\big)\right.\right.\right.$$

$$\left.\left.\left.\times \big((1 - s)[DX(T)](\sigma) + s[DX_h(T)](\sigma), F_h(T - \sigma)g(\sigma)\big)_{L_2^0}\right|\right]d\sigma$$

$$\leq C\left(1 + \left\|(1-s)X(T) + sX_h(T)\right\|_{L^p(\Omega;H)}^{p-2}\right)\left\|F_h(T - \cdot)g(\cdot)\right\|_{L^p(\Omega;L^2([0,T];L_2^0))}$$
$$\times \left(\left\|(1-s)DX(T) + sDX_h(T)\right\|_{L^p(\Omega;L^2([0,T];L_2^0))}\right),$$

which is uniformly bounded for all $s \in [0,1]$ by Theorems 5.7 and 5.8.

The proof of the error representation turns out to be a simple application of the mean-value theorem for Fréchet differentiable mappings and Bismut's integration by parts formula (4.20). Indeed, it holds that

$$\mathbf{E}\left[\varphi(X_h(T)) - \varphi(X(T))\right]$$

$$= \int_0^1 \mathbf{E}\left[\left(\varphi'(X(T) + s(X_h(T) - X(T))), X_h(T) - X(T)\right)\right] ds$$

$$= \int_0^1 \mathbf{E}\left[\left(\varphi'(X(T) + s(X_h(T) - X(T))), F_h(T)x_0 - \int_0^T F_h(T - \sigma)f(\sigma)\,d\sigma\right)\right] ds$$

$$+ \int_0^1 \mathbf{E}\left[\left(\varphi'(X(T) + s(X_h(T) - X(T))), \int_0^T F_h(T - \sigma)g(\sigma)\,dW(\sigma)\right)\right] ds.$$

By Lemma 4.7 and Proposition 4.12 we are allowed to apply (4.20) to the last summand. We obtain

$$\int_0^1 \mathbf{E}\left[\left(\varphi'(X(T) + s(X_h(T) - X(T))), \int_0^T F_h(T - \sigma)g(\sigma)\,dW(\sigma)\right)\right] ds$$

$$= \int_0^1 \int_0^T \mathbf{E}\left[\left([D\varphi'(X(T) + s(X_h(T) - X(T)))](\sigma), F_h(T - \sigma)g(\sigma)\right)_{L_2^0}\right] d\sigma\, ds$$

$$= \int_0^1 \int_0^T \mathbf{E}\left[\left(\varphi''(X(T) + s(X_h(T) - X(T)))\right.\right.$$

$$\times \left.\left.((1-s)[DX(T)](\sigma) + s[DX_h(T)](\sigma)), F_h(T - \sigma)g(\sigma)\right)_{L_2^0}\right] d\sigma\, ds.$$

This completes the proof.   □

In [45] the authors are concerned with stochastic evolution equations of the form

$$dX(t) + AX(t)\,dt = g\,dW(t), \quad \text{for } t \in [0,T], \tag{5.21}$$

where $-A$ is a generator of a strongly continuous semigroup $(E(t))_{t\geq 0}$ and g is a fixed operator in $L(U, H)$. If the covariance operator $Q: U \rightarrow U$ of the Wiener process satisfies the assumption

$$\int_0^T \|E(T - \sigma)g\|_{L_2^0}^2\,d\sigma = \text{Tr}\left(\int_0^T E(T - \sigma)gQg^*E(T - \sigma)^*\,d\sigma\right) < \infty$$

then the mild solution to (5.21) exists and is well-defined.

In order to formulate their error representation the authors consider the function
$u: H \times [0, T] \to \mathbb{R}$ which is given by

$$u(x, t) := \mathbf{E}[\varphi(Z(T; t, x))], \quad \text{for } x \in H, \ t \in [0, T],$$

and with

$$Z(T; t, x) = x + \int_t^T E(T - \sigma) g \, dW(\sigma).$$

Note that u solves an infinite dimensional *Kolmogorov's equation* associated to a
transformed version of the stochastic differential equation (5.21) (see [45, (3.4)]).
   For $\varphi \in C_b^2(H; \mathbb{R})$ the error representation in [45, Th. 3.1] is then written in the
following way:

$$\mathbf{E}[\varphi(X_h(T)) - \varphi(X(T))]$$

$$= \mathbf{E}[u(E_h(T) P_h x_0, 0) - u(E(T) x_0, 0)]$$

$$+ \frac{1}{2} \int_0^T \mathbf{E}\left[\left(u_{xx}(Y_h(t), t)(E_h(T - t) P_h + E(T - t)) g, F_h(T - t) g\right)_{L_2^0}\right] dt,$$

$$\text{(5.22)}$$

where the stochastic process $Y_h: [0, T] \times \Omega \to S_h$, $h \in (0, 1]$, is given by

$$Y_h(t) = E_h(T) P_h x_0 + \int_0^t E_h(T - \sigma) P_h g \, dW(\sigma), \quad t \in [0, T].$$

Applying the mean value theorem to the first summand in (5.22) yields

$$\mathbf{E}[u(E_h(T) P_h x_0, 0) - u(E(T) x_0, 0)]$$

$$= \int_0^1 \mathbf{E}\left[\left(\varphi'(((1 - s) E(T) + s E_h(T) P_h) x_0 + \int_0^T E(T - \sigma) g \, dW(\sigma)),\right.\right.$$

$$\left.\left. F_h(T) x_0\right)\right] ds.$$

In the same situation Theorem 5.9 gives the error representation

$$\mathbf{E}[\varphi(X_h(T)) - \varphi(X(T))]$$

$$= \int_0^1 \mathbf{E}\left[\left(\varphi'(X(T) + s(X_h(T) - X(T))), F_h(T) x_0\right)\right] ds$$

$$+ \int_0^1 \int_0^T \mathbf{E}\left[\left(\varphi''(X(T) + s(X_h(T) - X(T)))\right.\right.$$

$$\left.\left. \times ((1 - s) E(T - \sigma) + s E_h(T - \sigma) P_h) g, F_h(T - \sigma) g\right)_{L_2^0}\right] d\sigma \, ds,$$

$$\text{(5.23)}$$

where we applied Proposition 4.16 in order to obtain

$$[DX(T)](\sigma) = E(T - \sigma)g \quad \text{and} \quad [DX_h(T)](\sigma) = E_h(T - \sigma)P_h g$$

for all $\sigma \in [0, T]$.

Note that u_{xx} basically agrees with φ''. A comparison of (5.22) and (5.23) therefore shows that both representations are structurally very similar. Hence, it does not come as a surprise that an application of the arguments given in [45, Sects. 4, 5] to our error representation (5.23) also yields the same error estimates as in [45] for the heat equation, the linearized Cahn–Hilliard equation and the wave equation.

We close this section with a representation of the weak error of the spatio-temporal discretization. This time the error operator is denoted by

$$F_{kh}(t) := E_{kh}(t)P_h - E(t), \quad \text{for } t \in [0, T], \tag{5.24}$$

where E_{kh} is defined in (5.13). A comparable result, which again depends on u, is given in [46].

Theorem 5.10. *Under the assumptions of Theorem 5.9 the weak error (5.8) of the spatio-temporal discretization satisfies*

$$
\begin{aligned}
\mathbf{E}&\left[\varphi(X_h^{N_k}) - \varphi(X(T))\right] \\
&= \int_0^1 \mathbf{E}\left[\left(\varphi'(X(T) + s(X_h^{N_k} - X(T))), F_{kh}(T)x_0\right.\right. \\
&\qquad\qquad \left.- \int_0^T E_{kh}(T - \sigma)f_{kh}(\sigma) - E(T - \sigma)f(\sigma)\,\mathrm{d}\sigma\right)\bigg]\,\mathrm{d}s \\
&\quad + \int_0^1 \int_0^T \mathbf{E}\left[\left(\varphi''(X(T) + s(X_h^{N_k} - X(T)))\right.\right. \\
&\qquad\qquad \times \left((1 - s)[DX(T)](\sigma) + s[DX_h^{N_k}](\sigma)\right), \\
&\qquad\qquad \left. E_{kh}(T - \sigma)g_{kh}(\sigma) - E(T - \sigma)g(\sigma)\right)_{L_2^0}\bigg]\,\mathrm{d}\sigma\,\mathrm{d}s.
\end{aligned}
\tag{5.25}
$$

Proof. First we note that the stochastic process $[0, T] \ni t \mapsto E_{kh}(T - t)g_{kh}(t)$ is left-continuous and, hence, predictable. Therefore, Proposition 4.12 and Theorem 4.13 are still applicable and the proof is done in the same way as the proof of Theorem 5.9.   □

5.4   The Weak Error for the Linear Inhomogeneous Heat Equation

In this section we have a closer look on the weak discretization error of the linear heat equation with random inhomogeneities. In particular, we show that the weak order of convergence is almost twice the order of strong convergence.

For this recall from Theorems 3.10 and 3.14 that the schemes (5.6) and (5.7) are strongly convergent such that there exists a constant C, independent of $h, k \in (0, 1]$ such that

$$\|X_h(T) - X(T)\|_{L^p(\Omega; H)} \le C h$$

and

$$\|X_h^{N_k} - X(T)\|_{L^p(\Omega; H)} \le C(h + k^{\frac{1}{2}}).$$

By Assumption 5.4 the initial condition x_0 may be non-smooth. From this it follows that the constant C depends on $T^{-\frac{1}{2}}$. If it holds that $x_0 \in \dot{H}^1$ then the constant C is also independent of T.

In the analysis of the weak error of convergence, it turns out that we rely on essential parts of the proof of an estimate for the strong error. In particular, the following lemma can be seen as a special version of the estimate of the second summands in (3.22) and (3.57) but for $r = 1$.

Lemma 5.11. *Under Assumptions 3.3 and 5.1 and Assumption 5.2 with $\delta \in [\frac{1}{2}, 1]$, $p \in [2, \infty)$, there exists a constant C, such that*

$$\left\| \int_0^T E_h(T - \sigma) P_h f(\sigma) - E(T - \sigma) f(\sigma) \, d\sigma \right\|_{L^p(\Omega; H)} \le C h^2 \tag{5.26}$$

and

$$\left\| \int_0^T E_{kh}(T - \sigma) f_{kh}(\sigma) - E(T - \sigma) f(\sigma) \, d\sigma \right\|_{L^p(\Omega; H)} \le C \left(h^2 + k^\delta \right). \tag{5.27}$$

Proof. We first show (5.26). By the triangle inequality it holds that

$$\left\| \int_0^T E_h(T - \sigma) P_h f(\sigma) - E(T - \sigma) f(\sigma) \, d\sigma \right\|_{L^p(\Omega; H)}$$

$$\le \left\| \int_0^T F_h(T - \sigma)\big(f(\sigma) - f(T)\big) \, d\sigma \right\|_{L^p(\Omega; H)}$$

$$+ \left\| \int_0^T F_h(T - \sigma) f(T) \, d\sigma \right\|_{L^p(\Omega; H)}.$$

For the first summand we apply Lemma 3.8(i) with $\nu = 0$ and $\mu = 2$, and (5.9). Since $\delta > 0$ we obtain

$$\left\| \int_0^T F_h(T-\sigma)\big(f(\sigma)-f(T)\big)\,d\sigma \right\|_{L^p(\Omega;H)}$$

$$\leq Ch^2 \int_0^T (T-\sigma)^{-1}\|f(\sigma)-f(T)\|_{L^p(\Omega;H)}\,d\sigma \qquad (5.28)$$

$$\leq Ch^2 \int_0^T (T-\sigma)^{-1+\delta}\,d\sigma \leq Ch^2.$$

The second summand is estimated with Lemma 3.9(i) with $\rho = 0$ which yields

$$\left\| \int_0^T F_h(T-\sigma)f(T)\,d\sigma \right\|_{L^p(\Omega;H)} \leq Ch^2\|f(T)\|_{L^p(\Omega;H)} \leq Ch^2. \qquad (5.29)$$

Here we also used the fact that $\|f(T)\|_{L^p(\Omega;H)} < \infty$ from Assumption 5.2. Altogether, this completes the proof of (5.26).

The proof of (5.27) works in a similar way. First, we have

$$\left\| \int_0^T E_{kh}(T-\sigma)f_{kh}(\sigma) - E(T-\sigma)f(\sigma)\,d\sigma \right\|_{L^p(\Omega;H)}$$

$$\leq \left\| \int_0^T E_{kh}(T-\sigma)\big(f_{kh}(\sigma)-P_h f(\sigma)\big)\,d\sigma \right\|_{L^p(\Omega;H)}$$

$$+ \left\| \int_0^T F_{kh}(T-\sigma)\big(f(\sigma)-f(T)\big)\,d\sigma \right\|_{L^p(\Omega;H)}$$

$$+ \left\| \int_0^T F_{kh}(T-\sigma)f(T)\,d\sigma \right\|_{L^p(\Omega;H)}.$$

In order to estimate the first summand we apply (3.38) and obtain

$$\left\| E_{kh}(t)P_h x \right\| = \left\| R(kA_h)^j P_h x \right\| \leq \|x\| \qquad (5.30)$$

for all $x \in H, h,k \in (0,1], t \in [t_{j-1},t_j), j = 1,2,\dots$. Therefore, by (5.9)

$$\left\| \int_0^T E_{kh}(T-\sigma)\big(f_{kh}(\sigma)-P_h f(\sigma)\big)\,d\sigma \right\|_{L^p(\Omega;H)}$$

$$\leq \int_0^T \|f_{kh}(\sigma)-P_h f(\sigma)\|_{L^p(\Omega;H)}\,d\sigma \leq C \sum_{j=1}^{N_k+1}\int_{t_{j-1}}^{t_j}|t_{j-1}-\sigma|^\delta\,d\sigma \leq Ck^\delta.$$

This time by applying Lemma 3.12(i) with $\nu = 0$ and $\mu = 2$ the second summand is estimated in the same way as (5.28). We get

$$\left\| \int_0^T F_{kh}(T - \sigma)\big(f(\sigma) - f(T)\big)\,\mathrm{d}\sigma \right\|_{L^p(\Omega;H)} \le C\big(h^2 + k\big).$$

Finally, Lemma 3.13(i) yields

$$\left\| \int_0^T F_{kh}(T - \sigma)f(T)\,\mathrm{d}\sigma \right\|_{L^p(\Omega;H)} \le C\big(h^2 + k\big)\|f(T)\|_{L^p(\Omega;H)} \le C\big(h^2 + k\big)$$

which completes the proof.   □

Now, we are in a position to formulate the main result of this section.

Theorem 5.12. *Let Assumption 3.3 and Assumptions 5.1–5.6 with $\delta \in [\frac{1}{2}, 1]$, $p \in [2, \infty)$ hold. For a given $\varphi \in C_p^2(H; \mathbb{R})$ which satisfies (5.18)–(5.20) with the same $p \in [2, \infty)$ as above, the spatial semidiscrete approximation (5.6) and the spatio-temporal discretization (5.7) are weakly convergent. That is, there exists a constant C, independent of $h, k \in (0, 1]$, such that*

$$\big|\mathbf{E}\big[\varphi(X_h(T)) - \varphi(X(T))\big]\big| \le C(1 + |\log(h)|)\, h^2 \tag{5.31}$$

and

$$\big|\mathbf{E}\big[\varphi(X_h^{N_k}) - \varphi(X(T))\big]\big| \le C\big(1 + |\log(h)|\big)\big(h^2 + k^\delta\big). \tag{5.32}$$

For the proof of this theorem, the following stability estimates turn out to be essential. First, for a constant C, which is independent of $h, k \in (0, 1]$, it holds

$$\|X(T)\|_{L^p(\Omega;H)} + \|X_h(T)\|_{L^p(\Omega;H)} + \|X_h^{N_k}\|_{L^p(\Omega;H)} \le C \tag{5.33}$$

for all $h, k \in (0, 1]$, $p \in [2, \infty)$. Indeed, from Assumptions 5.1, 5.2 and 5.3 it directly follows that the norm of $X(T)$ is bounded. From (3.12) and (3.38) it holds that $\|E_h(t)\| \le C$ and $\|E_{kh}(t)\| \le C$, respectively. Therefore, by also applying Proposition 2.12 we obtain

$$\|X_h^{N_k}\|_{L^p(\Omega;H)}$$

$$\le \|x_0\| + \int_0^{t_{N_k}} \|f_{kh}(\sigma)\|_{L^p(\Omega;H)}\,\mathrm{d}\sigma + C\left(\mathbf{E}\left[\left(\int_0^{t_{N_k}} \|g_{kh}(\sigma)\|_{L_2^0}^2\,\mathrm{d}\sigma\right)^{\frac{p}{2}}\right]\right)^{\frac{1}{p}}$$

$$\le \|x_0\| + T \sup_{t\in[0,T]} \|f(t)\|_{L^p(\Omega;H)} + C\sqrt{T} \sup_{t\in[0,T]} \|g(t)\|_{L^p(\Omega;L_2^0)}.$$

The same bound holds for $\|X_h(T)\|$.

Secondly, the proof of Theorem 5.12 also depends on the following bounds of the Malliavin derivatives. Under the same assumptions as Theorem 5.12 there exists a constant $C > 0$, which is again independent of $h, k \in (0, 1]$, such that

$$\sup_{t \in [0,T]} \left\| [DX(T)](t) \right\|_{L^p(\Omega; L_2^0)} + \sup_{t \in [0,T]} \left\| [DX_h(T)](t) \right\|_{L^p(\Omega; L_2^0)}$$

$$+ \sup_{t \in [0,T]} \left\| [DX_h^{N_k}](t) \right\|_{L^p(\Omega; L_2^0)} \leq C, \tag{5.34}$$

for all $h, k \in (0, 1]$, $p \in [2, \infty)$. For this, Theorem 5.7 yields

$$\left\| [DX(T)](t) \right\|_{L^p(\Omega; L_2^0)} \leq \left\| \int_0^T E(T - \sigma) Df(\sigma, t; \cdot) \, d\sigma \right\|_{L^p(\Omega; L_2^0)}$$

$$+ \left\| E(T - t) g(t) \right\|_{L^p(\Omega; L_2^0)}$$

$$+ \left\| \int_0^T E(T - \sigma) Dg(\sigma, t; \cdot) \, dW(\sigma) \right\|_{L^p(\Omega; L_2^0)}.$$

Now, from Proposition 2.12 with $H = L_2^0$ we obtain

$$\left\| \int_0^T E(T - \sigma) Dg(\sigma, t; \cdot) \, dW(\sigma) \right\|_{L^p(\Omega; L_2^0)}$$

$$\leq C(p) \left(\mathbf{E} \left[\left(\int_0^T \| E(T - \sigma) Dg(\sigma, t; \cdot) \|_{L_2(U_0; L_2^0)}^2 \, d\sigma \right)^{\frac{p}{2}} \right] \right)^{\frac{1}{p}}$$

$$\leq C(p) \left(\int_0^T \| Dg(\sigma, t; \cdot) \|_{L^p(\Omega; L_2(U_0; L_2^0))}^2 \, d\sigma \right)^{\frac{1}{2}}$$

$$\leq C(p) \sqrt{T} \sup_{\sigma, t \in [0,T]} \| Dg(\sigma, t; \cdot) \|_{L^p(\Omega; L_2(U_0; L_2^0))}.$$

Thus,

$$\sup_{t \in [0,T]} \left\| [DX(T)](t) \right\|_{L^p(\Omega; L_2^0)}$$

$$\leq T \sup_{\sigma, t \in [0,T]} \| Df(\sigma, t; \cdot) \|_{L^p(\Omega; L_2^0)} + \sup_{t \in [0,T]} \| g(t) \|_{L^p(\Omega; L_2^0)}$$

$$+ C(p) \sqrt{T} \sup_{\sigma, t \in [0,T]} \| Dg(\sigma, t; \cdot) \|_{L^p(\Omega; L_2(U_0; L_2^0))},$$

which is bounded by Assumptions 5.3, 5.5 and 5.6. The same arguments can be applied to prove the bounds for $\left\| [DX_h(T)](t) \right\|_{L^p(\Omega; L_2^0)}$ and $\left\| [DX_h^{N_k}](t) \right\|_{L^p(\Omega; L_2^0)}$.

Proof (of Theorem 5.12). Starting point for the estimate of the weak error (5.31) of the semidiscrete approximation is Theorem 5.9. From there we get the error representation

$$\mathbf{E}\big[\varphi(X_h(T)) - \varphi(X(T))\big]$$

$$= \int_0^1 \mathbf{E}\Big[\Big(\varphi'\big(X(T) + s(X_h(T) - X(T))\big), F_h(T)x_0$$

$$- \int_0^T F_h(T - \sigma)f(\sigma)\,d\sigma\Big)\Big]\,ds$$

$$+ \int_0^1 \int_0^T \mathbf{E}\Big[\Big(\varphi''\big(X(T) + s(X_h(T) - X(T))\big)$$

$$\times \big((1 - s)[DX(T)](\sigma) + s[DX_h(T)](\sigma)\big), F_h(T - \sigma)g(\sigma)\Big)_{L_2^0}\Big]\,d\sigma\,ds.$$

We deal with the two summands separately. After taking the absolute value of the first summand, applying Hölder's inequality with exponents $r = \frac{p}{p-1}$ and $r' = p$ and (5.19) we derive the estimate

$$\Big|\int_0^1 \mathbf{E}\Big[\Big(\varphi'\big(X(T) + s(X_h(T) - X(T))\big), F_h(T)x_0 - \int_0^T F_h(T - \sigma)f(\sigma)\,d\sigma\Big)\Big]\,ds\Big|$$

$$\leq \int_0^1 \big\|\varphi'\big(X(T) + s(X_h(T) - X(T))\big)\big\|_{L^r(\Omega;H)}\,ds$$

$$\times \Big\|F_h(T)x_0 - \int_0^T F_h(T - \sigma)f(\sigma)\,d\sigma\Big\|_{L^{r'}(\Omega;H)}$$

$$\leq C\big(1 + \|X(T)\|_{L^p(\Omega;H)}^{p-1} + \|X_h(T)\|_{L^p(\Omega;H)}^{p-1}\big)$$

$$\times \Big(\|F_h(T)x_0\| + \Big\|\int_0^T F_h(T - \sigma)f(\sigma)\,d\sigma\Big\|_{L^p(\Omega;H)}\Big).$$

Further, an application of Lemma 3.8(i) with $\nu = 0$, $\mu = 2$ and (5.26) yields

$$\Big\|F_h(T)x_0\| + \Big\|\int_0^T F_h(T - \sigma)f(\sigma)\,d\sigma\Big\|_{L^p(\Omega;H)} \leq C\big(T^{-1}\|x_0\| + 1\big)h^2.$$

Hence, by also noting (5.33), the first summand is bounded by

$$\Big|\int_0^1 \mathbf{E}\Big[\Big(\varphi'\big(X(T) + s(X_h(T) - X(T))\big), F_h(T)x_0 - \int_0^T F_h(T - \sigma)f(\sigma)\,d\sigma\Big)\Big]\,ds\Big|$$

$$\leq C\big(1 + \|X(T)\|_{L^p(\Omega;H)}^{p-1} + \|X_h(T)\|_{L^p(\Omega;H)}^{p-1}\big)\big(T^{-1}\|x_0\| + 1\big)h^2 \leq Ch^2.$$

The proof of the second summand works as follows: After taking the absolute value, applying the Cauchy–Schwarz inequality and (5.20) we obtain

$$\left| \int_0^1 \int_0^T \mathbf{E}\Big[\big(\varphi''(X(T) + s(X_h(T) - X(T)))\right.$$

$$\left. \times \big((1 - s)[DX(T)](\sigma) + s[DX_h(T)](\sigma)\big), F_h(T - \sigma)g(\sigma)\big)_{L_2^0}\Big] d\sigma \, ds\right|$$

$$\leq \int_0^1 \int_0^T \mathbf{E}\Big[\big\|\varphi''(X(T) + s(X_h(T) - X(T)))\big\|_{L(H,H)}$$

$$\times \big\|(1 - s)[DX(T)](\sigma) + s[DX_h(T)](\sigma)\big\|_{L_2^0} \big\|F_h(T - \sigma)g(\sigma)\big\|_{L_2^0}\Big] d\sigma \, ds$$

$$\leq C \int_0^1 \int_0^T \mathbf{E}\Big[\big(1 + \big\|(1 - s)X(T) + sX_h(T)\big\|^{p-2}\big)$$

$$\times \big\|(1 - s)[DX(T)](\sigma) + s[DX_h(T)](\sigma)\big\|_{L_2^0} \big\|F_h(T - \sigma)g(\sigma)\big\|_{L_2^0}\Big] d\sigma \, ds.$$

Next, we apply the generalized Hölder inequality [2, Lemma 1.16] with exponents $r_1 = \frac{p}{p-2}, r_2 = r_3 = p$ to the integrand. Since $s \in [0, 1]$ we get

$$\mathbf{E}\Big[\big(1 + \big\|(1 - s)X(T) + sX_h(T)\big\|^{p-2}\big)$$

$$\times \big\|(1 - s)[DX(T)](\sigma) + s[DX_h(T)](\sigma)\big\|_{L_2^0} \big\|F_h(T - \sigma)g(\sigma)\big\|_{L_2^0}\Big]$$

$$\leq \Big(\mathbf{E}\Big[\big(1 + \big\|(1 - s)X(T) + sX_h(T)\big\|^{p-2}\big)^{r_1}\Big]\Big)^{\frac{1}{r_1}}$$

$$\times \big\|(1 - s)[DX(T)](\sigma) + s[DX_h(T)](\sigma)\big\|_{L^{r_2}(\Omega;L_2^0)}$$

$$\times \big\|F_h(T - \sigma)g(\sigma)\big\|_{L^{r_3}(\Omega;L_2^0)}$$

$$\leq C\Big(1 + \big\|X(T)\big\|_{L^p(\Omega;H)}^{p-2} + \big\|X_h(T)\big\|_{L^p(\Omega;L_2^0)}^{p-2}\Big) \big\|F_h(T - \sigma)g(\sigma)\big\|_{L^p(\Omega;L_2^0)}$$

$$\times \Big(\big\|[DX(T)](\sigma)\big\|_{L^p(\Omega;L_2^0)} + \big\|[DX_h(T)](\sigma)\big\|_{L^p(\Omega;L_2^0)}\Big).$$

By taking (5.33) and (5.34) into account we have proven the estimate

$$\left| \int_0^1 \int_0^T \mathbf{E}\Big[\big(\varphi''(X(T) + s(X_h(T) - X(T)))\right.$$

$$\left. \times \big((1 - s)[DX(T)](\sigma) + s[DX_h(T)](\sigma)\big), F_h(T - \sigma)g(\sigma)\big)_{L_2^0}\Big] d\sigma \, ds\right| \quad (5.35)$$

$$\leq C \int_0^T \big\|F_h(T - \sigma)g(\sigma)\big\|_{L^p(\Omega;L_2^0)} d\sigma.$$

As in the proof of [45, Theorem 4.1] the estimate is completed by splitting up the integral into two parts and applying Lemma 3.8(i) with $\mu = 2$, $\nu = 0$ to this first part and Lemma 3.8(i) with $\mu = \nu = 0$ to the second summand. Hence,

$$\int_0^T \left\| F_h(T - \sigma)g(\sigma) \right\|_{L^p(\Omega;L_2^0)} d\sigma$$

$$\leq C \left(h^2 \int_0^{T-h^2} (T - \sigma)^{-1} \|g(\sigma)\|_{L^p(\Omega;L_2^0)} d\sigma + \int_{T-h^2}^T \|g(\sigma)\|_{L^p(\Omega;L_2^0)} d\sigma \right)$$

$$\leq C h^2 \sup_{t \in [0,T]} \|g(t)\|_{L^p(\Omega;L_2^0)} \left(1 + \log(T) - 2\log(h)\right) \leq C\left(1 + |\log(h)|\right)h^2,$$

which completes the proof of (5.31).

In order to prove (5.32) we apply Theorem 5.10. As above we take the absolute value in (5.25) and estimate both summands separately. For the first summand it holds by Hölder's inequality with exponents $r = \frac{p}{p-1}$ and $r' = p$ and (5.19) that

$$\left| \int_0^1 \mathbf{E}\left[\left(\varphi'\left(X(T) + s(X_h^{N_k} - X(T))\right), \right. \right. $$

$$\left. \left. F_{kh}(T)x_0 - \int_0^T E_{kh}(T - \sigma)f_{kh}(\sigma) - E(T - \sigma)f(\sigma)\,d\sigma \right) \right] ds \right|$$

$$\leq \int_0^1 \left\| \varphi'\left(X(T) + s(X_h^{N_k} - X(T))\right) \right\|_{L^r(\Omega;H)} ds$$

$$\times \left\| F_{kh}(T)x_0 - \int_0^T E_{kh}(T - \sigma)f_{kh}(\sigma) - E(T - \sigma)f(\sigma)\,d\sigma \right\|_{L^{r'}(\Omega;H)}$$

$$\leq C\left(1 + \|X(T)\|_{L^p(\Omega;H)}^{p-1} + \|X_h^{N_k}\|_{L^p(\Omega;H)}^{p-1}\right)$$

$$\times \left(\|F_{kh}(T)x_0\| + \left\| \int_0^T E_{kh}(T - \sigma)f_{kh}(\sigma) - E(T - \sigma)f(\sigma)\,d\sigma \right\|_{L^p(\Omega;H)} \right)$$

$$\leq C\left(1 + \|X(T)\|_{L^p(\Omega;H)}^{p-1} + \|X_h^{N_k}\|_{L^p(\Omega;H)}^{p-1}\right)\left(T^{-1}\|x_0\| + 1\right)\left(h^2 + k^\delta\right),$$

where we applied Lemma 3.12(i) with $\mu = 2$, $\nu = 0$, and (5.27). By (5.33) the proof of the desired estimate of the first summand is complete.

It remains to estimate the absolute value of the second summand in (5.25). By exactly the same arguments as for (5.35) we obtain

$$\left| \int_0^1 \int_0^T \mathbf{E}\left[\left(\varphi''\left(X(T) + s(X_h^{N_k} - X(T))\right)\left((1-s)[DX(T)](\sigma) + s[DX_h^{N_k}](\sigma)\right), \right. \right. $$

$$\left. \left. E_{kh}(T - \sigma)g_{kh}(\sigma) - E(T - \sigma)g(\sigma) \right)_{L_2^0} \right] d\sigma\, ds \right|$$

$$\leq C \int_0^T \left\| E_{kh}(T - \sigma)g_{kh}(\sigma) - E(T - \sigma)g(\sigma) \right\|_{L^p(\Omega;L_2^0)} d\sigma.$$

Now, it holds

$$\int_0^T \left\| E_{kh}(T - \sigma) g_{kh}(\sigma) - E(T - \sigma) g(\sigma) \right\|_{L^p(\Omega; L_2^0)} d\sigma$$

$$\leq \int_0^T \left\| E_{kh}(T - \sigma) \big(g_{kh}(\sigma) - P_h g(\sigma) \big) \right\|_{L^p(\Omega; L_2^0)} d\sigma$$

$$+ \int_0^T \left\| F_{kh}(T - \sigma) g(\sigma) \right\|_{L^p(\Omega; L_2^0)} d\sigma$$

and we continue the estimate by applying (5.30) and (5.10) and we obtain

$$\int_0^T \left\| E_{kh}(T - \sigma) \big(g_{kh}(\sigma) - P_h g(\sigma) \big) \right\|_{L^p(\Omega; L_2^0)} d\sigma$$

$$\leq \int_0^T \left\| g_{kh}(\sigma) - P_h g(\sigma) \right\|_{L^p(\Omega; L_2^0)} d\sigma \leq C \sum_{j=1}^{N_k} \int_{t_{j-1}}^{t_j} |t_{j-1} - \sigma|^\delta d\sigma \leq C k^\delta.$$

Further, we also have by Assumption 5.3 and Lemma 3.12(i) with $\mu = 2$, $\nu = 0$ and $\mu = \nu = 0$, respectively,

$$\int_0^T \left\| F_{kh}(T - \sigma) g(\sigma) \right\|_{L^p(\Omega; L_2^0)} d\sigma$$

$$\leq \int_0^{T-h^2} \left\| F_{kh}(T - \sigma) g(\sigma) \right\|_{L^p(\Omega; L_2^0)} d\sigma + \int_{T-h^2}^T \left\| F_{kh}(T - \sigma) g(\sigma) \right\|_{L^p(\Omega; L_2^0)} d\sigma$$

$$\leq C \left((h^2 + k) \int_0^{T-h^2} (T - \sigma)^{-1} \| g(\sigma) \|_{L^p(\Omega; L_2^0)} d\sigma + \int_{T-h^2}^T \| g(\sigma) \|_{L^p(\Omega; L_2^0)} d\sigma \right)$$

$$\leq C \sup_{t \in [0,T]} \| g(t) \|_{L^p(\Omega; L_2^0)} \big(1 + \log(T) - 2 \log(h) \big) \big(h^2 + k \big)$$

$$\leq C (1 + |\log(h)|) \big(h^2 + k \big).$$

Putting all pieces together yields (5.32).   □

5.5 Some Generalizations

The most important generalization in this chapter is concerned with Assumption 5.1. As in our earlier chapters we assumed that A has a compact inverse. This condition is needed in order to apply the results from Chap. 3, but we only do so in Sect. 5.4.

As in [45, 46] the derivation of the error representation formula in Sect. 5.3 can be done under the much weaker assumption that $-A$ is only the generator of a strongly continuous semigroup. The same arguments as in [45, 46] can then be used to analyze the weak error of convergence for the stochastic wave equation in terms of our representation formula.

Further, under additional assumptions on the spatial discretization, for example if we consider the spectral Galerkin method from Example 3.7, it is possible to get rid of the $|\log(h)|$ singularity in the estimate of the weak error in Theorem 5.12. However, in our general situation it is still an open problem how to achieve the optimal order of convergence.

Chapter 6
Numerical Experiments

In this chapter we illustrate our theoretical findings for the spatially semidiscrete approximation through a series of numerical experiments. The details of the applied spatial discretization are developed in the first section. In particular, our experiments consider the stochastic heat equation with and without inhomogeneities in Sects. 6.2 and 6.3. Further, in Sects. 6.4 and 6.5 we are concerned with the geometric Brownian motion in infinite dimensions, again with and without inhomogeneities.

The usual approach to carry out numerical experiments for SPDEs is to first discretize the equation in space and time and then to estimate the strong and weak error through Monte-Carlo simulations. However, during the experiments it turned out that the weak error is in some situations so small, that it is completely dominated by the Monte-Carlo error for any reasonable number of sample paths.

To work around this problem, we applied a different strategy by using the Itô-isometry and Parseval's identity, which completely avoids the use of Monte-Carlo methods. The details for this are given in Sect. 6.2.

6.1 General Setting

Throughout this chapter we are in the following situation: We consider the separable Hilbert space $H = L^2([0, 1], \mathscr{B}([0, 1]), dx; \mathbb{R})$ and denote by $-A$ the Laplace operator with zero boundary conditions. As noted in Example 2.21, we have $\dot{H}^1 = H_0^1(0, 1)$ and $\dot{H}^2 = \mathrm{dom}(A) = H_0^1(0, 1) \cap H^2(0, 1)$. Further, the eigenvalues and eigenfunctions of A are given by

$$\lambda_j = j^2 \pi^2 \quad \text{and } e_j(y) = \sqrt{2}\sin(j\pi y) \text{ for all } j \in \mathbb{N}, \ j \geq 1, \ y \in [0, 1].$$

In all numerical experiments we are concerned with a spatial semidiscretization in terms of the standard finite element method from Example 3.6.

R. Kruse, *Strong and Weak Approximation of Semilinear Stochastic Evolution Equations*, Lecture Notes in Mathematics 2093, DOI 10.1007/978-3-319-02231-4_6, © Springer International Publishing Switzerland 2014

For this, let $N_h \in \mathbb{N}$ and $h := \frac{1}{N_h+1}$. We consider an equidistant partition of the spatial domain $[0,1]$ with $x_j := jh$, $j = 0,\ldots,N_h+1$. Then, for every h the spaces S_h are given as the set of all continuous functions on $[0,1]$, which are piecewise (affine) linear on all intervals $[x_{j-1}, x_j]$, $j = 1,\ldots,N_h+1$, and satisfy the zero boundary condition. From [53, Chap. 5.1] and [66, Chap. 1] we recall that $S_h \subset \dot{H}^1$ and $\dim(S_h) = N_h$. Further, the set $(\Phi_j)_{j=1}^{N_h}$ of so called *hat functions* or *pyramid functions* forms a basis of S_h, where Φ_j, $j \in \{1,\ldots,N_h\}$, is defined by

$$\Phi_j(x_i) = \begin{cases} 1, & \text{if } i = j, \\ 0, & \text{if } i \neq j. \end{cases}$$

Consequently, every $u_h \in S_h$ has the representation

$$u_h(x) = \sum_{j=1}^{\infty} u_h(x_j)\Phi_j(x), \quad \text{for all } x \in [0,1]. \tag{6.1}$$

In addition to the orthogonal projectors $P_h: \dot{H}^{-1} \to S_h$ and $R_h: \dot{H}^1 \to S_h$, which we introduced in Chap. 3, a given element $u \in \dot{H}^1$ may also be approximated by its *interpolant* $I_h u$. Here, the *interpolation operator* I_h is given by

$$I_h u = \sum_{j=1}^{\infty} u(x_j)\Phi_j. \tag{6.2}$$

Note, that by the Sobolev embedding theorem [2, 8.13 $\langle 2 \rangle$, S. 333], all elements in $\dot{H}^1 = H_0^1(0,1)$ have a continuous function in their equivalence class. Thus, the evaluation of $u \in \dot{H}^1$ at grid points is a well-defined operation if we say that we always evaluate the continuous representative of u.

From [12, Th. 3.1.5] and [33, Satz 91.6] it is well-known that I_h satisfies the following error estimates

$$\|I_h u - u\| \leq Ch^s \|u\|_s \quad \text{for all } u \in \dot{H}^s, \ s \in \{1,2\}, \ N_h \in \mathbb{N}. \tag{6.3}$$

Note that these estimates are comparable to Assumption 3.3.

Next, we recall from (3.11) that the discrete Laplacian is uniquely determined by the relationship

$$(u_h, v_h)_1 = (A_h u_h, v_h) \text{ for all } v_h \in S_h. \tag{6.4}$$

In particular, by inserting (6.1) and $v_h = \Phi_i$ for $i \in \{1,\ldots,N_h\}$ we obtain

$$(A_h u_h, \Phi_i) = (u_h, \Phi_i)_1 = \sum_{j=1}^{N_h} (\Phi_i, \Phi_j)_1 u_h(x_j).$$

Therefore, in terms of the basis $(\varPhi_j)_{j=1}^{N_h}$ the operator A_h is represented by the so called *stiffness matrix* with entries $a_{ij} = (\varPhi_i, \varPhi_j)_1$, for $i, j = 1, \ldots, N_h$. More precisely, the entries of the stiffness matrix are given by Hanke-Bourgeois [33, Bsp. 90.3]

$$a_{ij} = \int_0^1 \varPhi_i'(x)\varPhi_j'(x)\,dx = \begin{cases} \frac{2}{h}, & i = j, \\ -\frac{1}{h}, & |i-j| = 1, \\ 0, & |i-j| > 1. \end{cases}$$

That is, we obtain a well-known tridiagonal matrix of the form

$$(a_{ij})_{i,j=1}^{N_h} = \frac{1}{h}\begin{pmatrix} 2 & -1 & & & 0 \\ -1 & 2 & -1 & & \\ & -1 & 2 & & \\ & & & \ddots & -1 \\ 0 & & & -1 & 2 \end{pmatrix}.$$

The stiffness matrix is positive definite and symmetric (see [33, Prop. 90.4]) and its eigenvalues and eigenvectors are known to be [33, Aufg. VI.13]

$$\overline{\lambda}_{h,j} = \frac{4}{h}\left(\sin\left(\frac{j\pi}{2(N_h+1)}\right)\right)^2 \quad \text{and } s_{h,j} = \left(\sqrt{2}\sin\left(\frac{j\pi i}{N_h+1}\right)\right)_{i=1}^{N_h}$$

for every $j \in \{1, \ldots, N_h\}$. Since the eigenvalues are pairwise different, the eigenvectors $s_{h,j}$ are orthogonal. Now, if we define

$$e_{h,j} := \sum_{i=1}^{N_h} (s_{h,j})_i \varPhi_i,$$

we have the relationship

$$e_{h,j} = I_h e_j, \quad \text{for } j \in \{1, \ldots, N_h\}, \tag{6.5}$$

and it holds that

$$(e_{h,i}, e_{h,j})_1 = (A_h e_{h,i}, e_{h,j}) = \sum_{m,n=1}^{N_h} (s_{h,i})_m (s_{h,j})_n (A_h \varPhi_m, \varPhi_n)$$

$$= \sum_{m=1}^{N_h} (s_{h,i})_m \sum_{n=1}^{N_h} a_{mn}(s_{h,j})_n$$

$$= \sum_{m=1}^{N_h} (s_{h,i})_m \overline{\lambda}_{h,j}(s_{h,j})_m = \delta_{i,j}\overline{\lambda}_{h,j}\sum_{m=1}^{N_h}(s_{h,j})_m^2,$$

where $\delta_{i,j}$ denotes the Kronecker-δ. Hence, the set $(e_{h,j})_{j=1}^{N_h}$ is an orthogonal basis of S_h with respect to the inner product $(\cdot, \cdot)_1$. Moreover, by Hanke-Bourgeois [33, Prop. 55.3] it is true that

$$\sum_{m=1}^{N_h} (s_{h,j})_m^2 = N_h + 1 = \frac{1}{h}. \tag{6.6}$$

Further, it holds for $i, j \in \{1, \dots, N_h\}$

$$(\Phi_i, \Phi_j) = \begin{cases} \frac{2}{3}h, & i = j, \\ \frac{1}{6}h, & |i - j| = 1, \\ 0, & |i - j| > 1, \end{cases}$$

which are the entries of the so called *mass matrix*. Now, let us formally extend the eigenvectors $s_{h,j}$ to the boundary of the interval, that is $(s_{h,j})_0 = (s_{h,j})_{N_h+1} = 0$. Then, we use (6.6) and the fact that $2\sin(x)\cos(y) = \sin(x - y) + \sin(x + y)$ and we obtain

$$(e_{h,i}, e_{h,j})$$

$$= \sum_{m,n=1}^{N_h} (s_{h,i})_m (s_{h,j})_n (\Phi_m, \Phi_n)$$

$$= \sum_{m=1}^{N_h} (s_{h,i})_m \left(\frac{2}{3} h (s_{h,j})_m + \frac{1}{6} h \left((s_{h,j})_{m-1} + (s_{h,j})_{m+1} \right) \right)$$

$$= h \sum_{m=1}^{N_h} (s_{h,i})_m (s_{h,j})_m \left(\frac{2}{3} + \frac{1}{3} \cos \left(\frac{j\pi}{N_h + 1} \right) \right) = \delta_{i,j} \left(\frac{2}{3} + \frac{1}{3} \cos \left(\frac{j\pi}{N_h + 1} \right) \right).$$
$$\tag{6.7}$$

Thus, the functions $(e_{h,j})_{j=1}^{N_h}$ are also orthogonal with respect to the inner product in H. For $i \in \{1, \dots, N_h\}$ let us define

$$\lambda_{h,i} := \frac{1}{h} \bar{\lambda}_{h,i} \left(\frac{2}{3} + \frac{1}{3} \cos \left(\frac{i\pi}{N_h + 1} \right) \right)^{-1},$$

then we have finally shown

$$(A_h e_{h,i}, e_{h,j}) = \delta_{i,j} \lambda_{h,i} (e_{h,i}, e_{h,j}) \quad \text{for all } i, j \in \{1, \dots, N_h\}.$$

Therefore, the orthogonal functions $(e_{h,j})_{j=1}^{N_h}$ are indeed a basis of eigenfunctions of A_h in S_h with corresponding eigenvalues $\lambda_{h,j}, j \in \{1, \dots, N_h\}$.

We close this section with the following observation: For arbitrary $j \in \mathbb{N}$ we write $j = 2m(N_h + 1) + n$ with $n \in \{0, \ldots, 2N_h + 2\}$ and $m \in \mathbb{N}$ and obtain by the periodicity of the sine function that

$$(s_{h,j})_i = e_j(x_i) = \sqrt{2} \sin \left(\frac{(n + 2m(N_h + 1))\pi i}{N_h + 1} \right) = e_n(x_i) = (s_{h,n})_i .$$

In particular, if $j = m(N_h + 1)$ for some $m \in \mathbb{N}$, that is $n \in \{0, N_h + 1\}$, we have $e_j(x_i) = 0$ for all $i \in \{1, \ldots, N_h\}$.

Further, if $n \in \{N_h + 2, \ldots, 2N_h + 1\}$ we apply the fact that $\sin(x) = \sin(\pi - x)$. This yields

$$e_n(x_i) = \sqrt{2} \sin \left(\frac{n\pi i}{N_h + 1} \right) = \sqrt{2} \sin \left(\pi - \frac{n\pi i}{N_h + 1} \right)$$

$$= -\sqrt{2} \sin \left(2i\pi - \frac{n\pi i}{N_h + 1} \right) = -\sqrt{2} \sin \left(\frac{(2(N_h + 1) - n)i\pi}{N_h + 1} \right)$$

$$= e_\ell(x_i)$$

with $\ell = 2(N_h + 1) - n \in \{1, \ldots, N_h + 1\}$.

Altogether, this proves for $j = 2m(N_h + 1) + n$

$$I_h e_j = \begin{cases} e_{h,n}, & \text{if } n \in \{1, \ldots, N_h\}, \\ -e_{h,\ell}, & \text{if } n \in \{N_h + 2, \ldots, 2N_h + 1\}, \ \ell = 2(N_h + 1) - n, \\ 0, & \text{if } n \in \{0, N_h + 1\}. \end{cases} \quad (6.8)$$

This generalizes the relationship (6.5).

In signal and image processing this relationship is the reason for the so called *aliasing effect*. After the discretization of a continuous signal, higher and lower frequencies are represented by the same single basis element $e_{h,j}$ in S_h. This may cause two different signals to be represented by the same discrete *alias*. Hence, it is impossible to reconstruct the original frequencies from the discretized signal which sometimes leads to strange effects in computer graphics, for example Moiré patterns. For more details we refer to the discussion in [33, S. 401].

6.2 Stochastic Heat Equation with Additive Noise

Our first numerical experiment is concerned with the semidiscretization of the homogeneous stochastic heat equation with additive noise from Example 2.21.

More precisely, our aim is to compute the strong and weak error of the spatially semidiscrete approximation of the mild solution to

$$dX(t) + AX(t)\,dt = dW(t), \quad \text{for all } t \in [0, T],$$

$$X(0) = X_0. \tag{6.9}$$

Since the numerical treatment of the initial condition is more or less the same as in deterministic examples, we set it equal to zero, that is $X_0 \equiv 0 \in H$. Thus, the mild solution to (6.9) is given by the stochastic convolution

$$X(t) = \int_0^t E(t - \sigma)\,dW(\sigma), \quad t \in [0, T].$$

Concerning the Wiener process W we assume that its covariance operator Q diagonalizes with respect to the same eigenfunction basis as A, that is

$$Qe_j = \mu_j e_j \quad \text{for } j \in \mathbb{N},$$

where $\mu_j \geq 0$ denotes the j-th eigenvalue of Q.

In Chaps. 3 and 5 we analyzed the strong error and the weak error of convergence of spatial semidiscretization $X_h : [0, T] \times \Omega \to S_h$, which solves the stochastic evolution equation

$$dX_h(t) + A_h X_h(t)\,dt = P_h\,dW(t), \quad \text{for } 0 \leq t \leq T,$$

$$X_h(0) = 0, \tag{6.10}$$

where P_h denotes the generalized orthogonal projector onto S_h and $A_h : S_h \to S_h$ is the discrete Laplacian as introduced in Sect. 6.1.

We are interested in the strong error at the final time $t = T$, that is

$$e_h^s(T) := \left(\mathbf{E}\left[\| X_h(T) - X(T) \|^2 \right] \right)^{\frac{1}{2}}$$

and the weak error

$$e_h^w(T) := \left| \mathbf{E}\left[\varphi(X_h(T)) - \varphi(X(T)) \right] \right|,$$

where we choose the same test function

$$\varphi(h) := \| h \|^2$$

for all experiments in this chapter.

Our choice of φ and the overall setting is motivated by the fact that we want to avoid costly Monte-Carlo simulations. Instead, we use Parseval's identity and the Itô-isometry in order to directly determine a deterministic series representation of both errors.

First, it holds

$$
\mathbf{E}\big[\|X(T)\|^2\big] = \int_0^T \|E(T - \sigma)\|_{L_2^0}^2 \, d\sigma = \sum_{j=1}^{\infty} \mu_j \int_0^T e^{-2\lambda_j(T-\sigma)} \, d\sigma
$$

$$
= \sum_{j=1}^{\infty} \frac{\mu_j}{2\lambda_j}\big(1 - e^{-2\lambda_j T}\big).
$$

(6.11)

Since the eigenvalues λ_j of A are known and if we neglect the cut-off error, we can numerically compute this series for every choice of μ_j, $j \in \mathbb{N}$.

On the other hand, we have

$$
\mathbf{E}\big[\|X_h(T)\|^2\big] = \int_0^T \|E_h(T - \sigma)P_h\|_{L_2^0}^2 \, d\sigma = \sum_{j=1}^{\infty} \mu_j \int_0^T \|E_h(T - \sigma)P_h e_j\|^2 \, d\sigma.
$$

Here, the computation is not so straightforward, since $(P_h e_j)_{j=1}^{N_h}$ is not a system of eigenfunctions of A_h. By recalling our results from the previous section it is computationally more convenient to consider

$$
\overline{X}_h(T) = \int_0^T E_h(T - \sigma)J_h \, dW(\sigma)
$$

in place of $X_h(T)$. That is, in (6.10) we replace the orthogonal projector P_h by a finitely based variant of the interpolation operator (6.2). Here, $J_h \colon \dot{H}^1 \to S_h$ is defined by

$$
J_h e_j := \begin{cases} I_h e_j, & \text{for } j \in \{1, \ldots, N_h\}, \\ 0, & \text{for } j \geq N_h + 1. \end{cases}
$$

(6.12)

For a justification we refer to Lemma 6.1 below.

In this case we obtain

$$
\mathbf{E}\big[\|\overline{X}_h(T)\|^2\big] = \int_0^T \|E_h(T - \sigma)J_h\|_{L_2^0}^2 \, d\sigma = \sum_{j=1}^{N_h} \mu_j \int_0^T \|E_h(T - \sigma)I_h e_j\|^2 \, d\sigma
$$

$$
= \sum_{j=1}^{N_h} \frac{\mu_j}{2\lambda_{h,j}}\big(1 - e^{-2\lambda_{h,j} T}\big)\|e_{h,j}\|^2,
$$

(6.13)

which can now be computed explicitly by using (6.7). Thus, we are in a position to perform a computation of the weak error.

Table 6.1 Homogeneous stochastic heat equation

N_h	Strong error	EOC	Weak error	EOC
250	2.335946e−03		1.030124e−04	
500	1.149847e−03	1.022564	2.845954e−05	1.855834
1,000	5.654677e−04	1.023926	7.759938e−06	1.874795
2,000	2.779735e−04	1.024497	2.094525e−06	1.889422
4,000	1.366221e−04	1.024757	5.607766e−07	1.901125
8,000	6.714305e−05	1.024880	1.491400e−07	1.910759
16,000	3.299615e−05	1.024940	3.944194e−08	1.918864

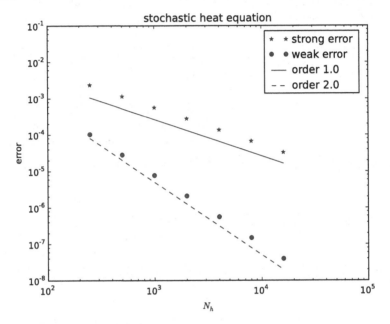

Fig. 6.1 Strong error (marked by *asterisks*) and weak error (marked by *dots*) for the semidiscretization of the homogeneous stochastic heat equation (values from Table 6.1)

Our results are shown on the right hand side of Table 6.1 and by the dotted line in Fig. 6.1. To be more precise, we set $T = 1$ and $\mu_j := 10\left(\frac{1}{j}\right)^{1.05}$, $j \in \mathbb{N}$, and, for each choice of N_h, we take the first $\frac{N_h^2}{10}$ summands of the infinite series in (6.11). Then, the weak error is approximated by the absolute value of the difference of (6.11) and (6.13), that is

$$e_h^w(T) \approx \left| \sum_{j=1}^{N_h} \frac{\mu_j}{2\lambda_{h,j}} \left(1 - e^{-2\lambda_{h,j}T}\right) \|e_{h,j}\|^2 - \sum_{j=1}^{N_h^2/10} \frac{\mu_j}{2\lambda_j} \left(1 - e^{-2\lambda_j T}\right) \right|.$$

The last column of Table 6.1 contains the *experimental order of convergence* (EOC) of the weak error, that is for two successive values of N_h, say N_{h_1} and N_{h_2}, we compute

$$\mathrm{EOC}(N_{h_1}, N_{h_2}) := \frac{\log\left(e_{h_2}^w(T)\right) - \log\left(e_{h_1}^w(T)\right)}{\log(N_{h_1}) - \log(N_{h_2})}. \tag{6.14}$$

In fact, it holds that $-\mathrm{EOC}(N_{h_1}, N_{h_2})$ is the slope of a line connecting the points $(N_{h_1}, e_{h_1}^w(T))$ and $(N_{h_2}, e_{h_2}^w(T))$ in a figure with logarithmic scaled axes such as Fig. 6.1 and, therefore, it is an experimental estimate of the order of convergence.

Next, we also approximate the strong error of convergence. We again replace $X_h(T)$ by $\overline{X}_h(T)$ and we get

$$\|X_h(T) - X(T)\|_{L^2(\Omega;H)} \le \|X_h(T) - \overline{X}_h(T)\|_{L^2(\Omega;H)} + \|\overline{X}_h(T) - X(T)\|_{L^2(\Omega;H)}.$$

The first summand is analyzed in Lemma 6.1 and we concentrate on the second summand. Let us write

$$X(t) = \sum_{j=1}^{\infty} \sqrt{\mu_j} \int_0^T e^{-\lambda_j(T-\sigma)} \,\mathrm{d}\beta_j(\sigma)\, e_j$$

$$= \sum_{j=1}^{N_h} \sqrt{\mu_j} \int_0^T e^{-\lambda_j(T-\sigma)} \,\mathrm{d}\beta_j(\sigma)\, e_j + \sum_{j=N_h+1}^{\infty} \sqrt{\mu_j} \int_0^T e^{-\lambda_j(T-\sigma)} \,\mathrm{d}\beta_j(\sigma)$$

$$=: X_{1,N_h}(T) + X_{N_h+1,\infty}(T).$$

Hence, it holds

$$\|\overline{X}_h(T) - X(T)\|_{L^2(\Omega;H)}$$
$$\le \|\overline{X}_h(T) - J_h X_{1,N_h}(T)\|_{L^2(\Omega;H)} + \|J_h X_{1,N_h}(T) - X_{1,N_h}(T)\|_{L^2(\Omega;H)}$$
$$+ \|X_{N_h+1,\infty}(T)\|_{L^2(\Omega;H)}.$$

First, we use (6.3) and obtain

$$\|(J_h - \mathrm{Id}_H)X_{1,N_h}(T)\|_{L^2(\Omega;H)} = \|(I_h - \mathrm{Id}_H)X_{1,N_h}(T)\|_{L^2(\Omega;H)}$$
$$\le Ch^{1+r}\|X_{1,N_h}(T)\|_{L^2(\Omega;\dot{H}^{1+r})},$$

which is why we also neglect this summand in the numerical experiment. For the first summand we recall the relationship (6.5) and get

$$\|\overline{X}_h(T) - J_h X_{1,N_h}(T)\|^2_{L^2(\Omega;H)}$$

$$= \sum_{j=1}^{N_h} \mu_j \, \mathbf{E}\left[\left| \int_0^T e^{-\lambda_{h,j}(T-\sigma)} - e^{-\lambda_j(T-\sigma)} \, d\beta_j(\sigma) \right|^2\right] \|e_{h,j}\|^2 \tag{6.15}$$

$$= \sum_{j=1}^{N_h} \mu_j \int_0^T \left(e^{-\lambda_{h,j}(T-\sigma)} - e^{-\lambda_j(T-\sigma)}\right)^2 d\sigma \|e_{h,j}\|^2.$$

Recalling (6.7) and since

$$\int_0^T \left(e^{-\lambda_{h,j}(T-\sigma)} - e^{-\lambda_j(T-\sigma)}\right)^2 d\sigma$$

$$= \frac{1}{2\lambda_{h,j}}\left(1 - e^{-2\lambda_{h,j}T}\right) - \frac{2}{\lambda_{h,j} + \lambda_j}\left(1 - e^{-(\lambda_{h,j}+\lambda_j)T}\right) + \frac{1}{2\lambda_j}\left(1 - e^{-2\lambda_j T}\right),$$

this summand can be computed explicitly. Moreover, we approximate the third summand by

$$\|X_{N_h+1,\infty}(T)\|^2_{L^2(\Omega;H)} = \sum_{j=N_h+1}^{\infty} \frac{\mu_j}{2\lambda_j}\left(1 - e^{-2\lambda_j T}\right)$$

$$\approx \sum_{j=N_h+1}^{N_h^2/10} \frac{\mu_j}{2\lambda_j}\left(1 - e^{-2\lambda_j T}\right). \tag{6.16}$$

The computed sum of the two terms (6.15) and (6.16) is shown in Table 6.1 and Fig. 6.1 and called the strong error. For the computation of the strong error we use the same values for the eigenvalues μ_j as for the weak error.

We remark that both computational approximations of the strong and weak error are in line with our theoretical findings. By our choice of the eigenvalues μ_j of Q it holds that $\|g\|_{L^0_{2,r}} < \infty$ for every $r \in [0, 0.025)$. In particular, by Theorem 3.10 we expect a strong order of convergence of slightly below 1.025. Similarly, the assumptions of Theorem 5.12 are satisfied as well and we expect a weak order of convergence of order slightly less than 2. Especially for higher values of N_h we find this behaviour confirmed by our numerical experiments. For lower values of N_h it appears that we systematically underestimate the weak error which results in lower values of the experimental order of convergence.

For the rest of this section, we return to the question, if the replacement of the orthogonal projector P_h by the finitely based variant J_h of the interpolation operator I_h is justified. In practice, it is quite common to use the interpolation operator, especially in our situation, where $I_h e_j$ opens the door to the fast algorithms of the *discrete sine transformation* (DST), which are based on the same idea as the *fast Fourier transform* (FFT) algorithms.

However, to the best of our knowledge, it is only barely analyzed in the literature, how the replacement of P_h by J_h affects the order of convergence. At least in our situation the following lemma shows that the use of J_h leaves the order of strong convergence unchanged. Although we will use this replacement in all numerical examples in this chapter, we leave a rigorous justification in a more general situation as a subject to future research.

Lemma 6.1. *Under the condition that* $\mathrm{Tr}(Q) < \infty$, *there exists a constant* C *independent of* N_h *such that*

$$\left\| X_h(T) - \overline{X}_h(T) \right\|_{L^2(\Omega;H)} \leq Ch \quad \text{for all } N_h \in \mathbb{N}.$$

Proof. First, it holds by the Itô-isometry

$$\left\| X_h(T) - \overline{X}_h(T) \right\|_{L^2(\Omega;H)}^2 = \sum_{j=1}^{\infty} \mu_j \int_0^T \left\| E_h(T-\sigma)(P_h - J_h)e_j \right\|^2 d\sigma.$$

Next, we recall from (3.15) the fact that $\| A_h^{-\frac{1}{2}} P_h x \| \leq \|x\|_{-1}$ for all $x \in \dot{H}^{-1}$. By Lemma B.9(*iii*) with $\rho = 1$ we obtain

$$\int_0^T \left\| E_h(T-\sigma)(P_h - J_h)e_j \right\|^2 d\sigma$$

$$= \int_0^T \left\| A_h^{\frac{1}{2}} E_h(T-\sigma) A_h^{-\frac{1}{2}} P_h(\mathrm{Id}_H - J_h)e_j \right\|^2 d\sigma$$

$$\leq C \left\| A_h^{-\frac{1}{2}} P_h(\mathrm{Id}_H - J_h)e_j \right\|^2 \leq C \left\| (\mathrm{Id}_H - J_h)e_j \right\|_{-1}^2.$$

Since $Je_j = 0$ and $\lambda_j \leq Ch^2$ for $j \geq N_h + 1$, we get

$$\int_0^T \left\| E_h(T-\sigma)(P_h - J_h)e_j \right\|^2 d\sigma$$

$$\leq C \left\| (\mathrm{Id}_H - J_h)e_j \right\|_{-1}^2 = C \|e_j\|_{-1}^2 = C\lambda_j^{-1} \leq Ch^2$$

for all $j \geq N_h + 1$. Hence,

$$\left\| X_h(T) - \overline{X}_h(T) \right\|_{L^2(\Omega;H)}^2 \leq C \sum_{j=1}^{N_h} \mu_j \left\| (\mathrm{Id}_H - J_h)e_j \right\|_{-1}^2 + Ch^2 \sum_{j=N_h+1}^{\infty} \mu_j$$

$$\leq C \sum_{j=1}^{N_h} \mu_j \left\| (\mathrm{Id}_H - J_h)e_j \right\|_{-1}^2 + Ch^2 \mathrm{Tr}(Q).$$

For the first sum we recall the definition of the $\|\cdot\|_{-1}$-norm. Then, it holds

$$\sum_{j=1}^{N_h} \mu_j \|(\mathrm{Id}_H - \mathrm{J}_h)e_j\|_{-1}^2$$

$$= \sum_{j=1}^{N_h} \mu_j \sum_{i=1}^{\infty} \lambda_i^{-1} \big((\mathrm{Id}_H - \mathrm{J}_h)e_j, e_i\big)^2$$

$$= \sum_{j=1}^{N_h} \mu_j \sum_{i=1}^{N_h} \lambda_i^{-1} \big((\mathrm{Id}_H - \mathrm{J}_h)e_j, e_i\big)^2 + \sum_{j=1}^{N_h} \mu_j \sum_{i=N_h+1}^{\infty} \lambda_i^{-1} \big((\mathrm{Id}_H - \mathrm{J}_h)e_j, e_i\big)^2.$$

As above, since $\lambda_i^{-1} \le C h^2$ for $i \ge N_h + 1$, we estimate the second sum by

$$\sum_{j=1}^{N_h} \mu_j \sum_{i=N_h+1}^{\infty} \lambda_i^{-1} \big((\mathrm{Id}_H - \mathrm{J}_h)e_j, e_i\big)^2 \le C h^2 \sum_{j=1}^{N_h} \mu_j \sum_{i=N_h+1}^{\infty} \big((\mathrm{Id}_H - \mathrm{J}_h)e_j, e_i\big)^2$$

$$= C h^2 \sum_{j=1}^{N_h} \mu_j \sum_{i=N_h+1}^{\infty} \big(\mathrm{J}_h e_j, e_i\big)^2$$

$$\le C h^2 \sum_{j=1}^{N_h} \mu_j \|\mathrm{J}_h e_j\|^2.$$

Therefore, since by (6.7) we have $\|\mathrm{J}_h e_j\|^2 = \|\mathrm{I}_h e_j\|^2 \le 1$, it follows

$$\sum_{j=1}^{N_h} \mu_j \sum_{i=N_h+1}^{\infty} \lambda_i^{-1} \big((\mathrm{Id}_H - \mathrm{J}_h)e_j, e_i\big)^2 \le C h^2 \mathrm{Tr}(Q)$$

and the proof of the lemma is complete, if we derive a similar estimate for

$$\sum_{j=1}^{N_h} \mu_j \sum_{i=1}^{N_h} \lambda_i^{-1} \big((\mathrm{Id}_H - \mathrm{J}_h)e_j, e_i\big)^2$$

$$= \sum_{j=1}^{N_h} \mu_j \lambda_j^{-1} \big((\mathrm{Id}_H - \mathrm{J}_h)e_j, e_i\big)^2 + \sum_{j=1}^{N_h} \mu_j \sum_{\substack{i=1 \\ i \ne j}}^{N_h} \lambda_i^{-1} \big((\mathrm{Id}_H - \mathrm{J}_h)e_j, e_i\big)^2. \tag{6.17}$$

For the first sum in (6.17) we obtain by the Cauchy–Schwarz inequality

$$\sum_{j=1}^{N_h} \mu_j \lambda_j^{-1} \big((\mathrm{Id}_H - \mathrm{J}_h)e_j, e_j\big)^2 \leq C \sum_{j=1}^{N_h} \mu_j \lambda_j^{-1} \big\|(\mathrm{Id}_H - \mathrm{J}_h)e_j\big\|^2$$

$$\leq Ch^2 \sum_{j=1}^{N_h} \mu_j = Ch^2 \mathrm{Tr}(Q),$$

where we also applied the estimate (6.3) which yields

$$\big\|(\mathrm{Id}_H - \mathrm{J}_h)e_j\big\| = \big\|(\mathrm{Id}_H - \mathrm{I}_h)e_j\big\| \leq Ch\lambda_j^{\frac{1}{2}}.$$

For the second sum in (6.17) we get

$$\big((\mathrm{Id}_H - \mathrm{J}_h)e_j, e_i\big)^2 = \big(\mathrm{I}_h e_j, -e_i\big)^2 = \big(\mathrm{I}_h e_j, \mathrm{I}_h e_i - e_i\big)^2$$

by the orthogonality of the families $(e_j)_{j \in \mathbb{N}}$ and $(\mathrm{I}_h e_j)_{j=1}^{N_h} = (e_{h,j})_{j=1}^{N_h}$. Hence,

$$\sum_{j=1}^{N_h} \mu_j \sum_{\substack{i=1 \\ i \neq j}}^{N_h} \lambda_i^{-1} \big((\mathrm{Id}_H - \mathrm{J}_h)e_j, e_i\big)^2$$

$$= \sum_{j=1}^{N_h} \mu_j \sum_{\substack{i=1 \\ i \neq j}}^{N_h} \lambda_i^{-1} \big(e_{h,j}, (\mathrm{I}_h - \mathrm{Id}_H)e_i\big)^2$$

$$= \sum_{j=1}^{N_h} \mu_j \sum_{\substack{i=1 \\ i \neq j}}^{N_h} \big(e_{h,j}, (\mathrm{I}_h - \mathrm{Id}_H)A^{-\frac{1}{2}}e_i\big)^2$$

$$= \sum_{i=1}^{N_h} \sum_{\substack{j=1 \\ j \neq i}}^{N_h} \big(\frac{1}{\|e_{h,j}\|}e_{h,j}, Q_h^{\frac{1}{2}} P_h(\mathrm{I}_h - \mathrm{Id}_H)A^{-\frac{1}{2}}e_i\big)^2,$$

where the positive definite and self-adjoint operator $Q_h: S_h \to S_h$ is given by

$$Q_h e_{h,n} := \mu_n \|e_{h,n}\|^2 e_{h,n} \quad \text{for all } n \in \{1, \ldots, N_h\}.$$

A further application of Parseval's identity yields

$$\sum_{i=1}^{N_h} \sum_{\substack{j=1 \\ j \neq i}}^{N_h} \big(\frac{1}{\|e_{h,j}\|}e_{h,j}, Q_h^{\frac{1}{2}} P_h(\mathrm{I}_h - \mathrm{Id}_H)A^{-\frac{1}{2}}e_i\big)^2 \leq \sum_{i=1}^{N_h} \big\|Q_h^{\frac{1}{2}} P_h(\mathrm{I}_h - \mathrm{Id}_H)A^{-\frac{1}{2}}e_i\big\|^2$$

$$\leq \big\|Q_h^{\frac{1}{2}} P_h(\mathrm{I}_h - \mathrm{Id}_H)A^{-\frac{1}{2}}\big\|_{L_2(H, S_h)}^2.$$

The estimate is completed by an application of Proposition 2.9(*iii*), which gives

$$\left\| Q_h^{\frac{1}{2}} P_h (I_h - \mathrm{Id}_H) A^{-\frac{1}{2}} \right\|_{L_2(H,S_h)}^2 \leq \left\| Q_h^{\frac{1}{2}} \right\|_{L_2(S_h,S_h)}^2 \left\| P_h (\mathrm{Id}_H - I_h) A^{-\frac{1}{2}} \right\|_{L(H,S_h)}^2$$

$$\leq Ch^2 \sum_{n=1}^{\infty} \mu_n \| e_{h,n} \|^2 \leq Ch^2 \mathrm{Tr}(Q),$$

where we also used that $\| P_h (\mathrm{Id}_H - I_h) A^{-\frac{1}{2}} u \| \leq Ch \| A^{-\frac{1}{2}} u \|_1 = Ch \| u \|$ by (6.3) and $\| e_{h,n} \| \leq 1$ by (6.7). This completes the proof. $\square$

Remark 6.2. The decision to use the finitely based interpolation operator J_h instead of I_h is due to the aliasing effect. Let us observe that for arbitrary $u \in \dot{H}^1$ we have

$$I_h u = \sum_{j=1}^{\infty} (u, e_j) I_h e_j = \sum_{n=1}^{N_h} \underbrace{\sum_{j=1}^{\infty}}_{I_h e_j = \pm e_{h,n}} (u, e_j) e_{h,n},$$

where we used the relationship (6.8). Consequently, each Fourier coefficient of $I_h u$ in terms of the basis $(e_{h,n})_{n=1}^{N_h}$ is, in general, an infinite series which we need to cut-off at some point during a numerical computation. In principle, the cut-off of the series is equivalent to considering a finitely based variant of I_h in the first place.

However, while this argument still leaves some room for the definition of J_h our particular choice is also important for Lemma 6.1. Here, it is unclear at the beginning of the proof how to estimate the summands $j \geq N_h + 1$ if $J_h e_j \neq 0$.

6.3 Inhomogeneous Stochastic Heat Equation

Our second numerical experiment is concerned with the inhomogeneous stochastic heat equation. That is, we consider the stochastic evolution equation on $H = L^2([0,1]; \mathbb{R})$

$$dX(t) + \left[AX(t) + f(t) \right] dt = (A + \mathrm{Id}_H) \, dW(t), \quad \text{for all } t \in [0,T], \tag{6.18}$$
$$X(0) = 0,$$

where A and W are the same as in the previous section and $T = 1$. This time we choose the eigenvalues of the covariance operator Q to be $\mu_j := 10 \left(\frac{1}{j} \right)^{5.05}$ and, as our main novelty in this experiment, we introduce a stochastic process $f : [0,T] \times \Omega \to H$ as inhomogeneity.

To be more precise, let f be given by $f(t) = -AW(t)$, which is a well-defined H-valued predictable process, since

$$\mathbf{E} \left[\| AW(t) \|^2 \right] = \sum_{j=1}^{\infty} \mu_j \lambda_j^2 t = 10 \pi^4 t \sum_{j=1}^{\infty} \frac{j^4}{j^{5.05}} < \infty.$$

Then, the mild solution to (6.18) is given by

$$X(t) = -\int_0^t E(T - \sigma)f(\sigma)\,d\sigma + \int_0^t E(t - \sigma)(A + \mathrm{Id}_H)\,dW(\sigma)$$

$$= \int_0^t E(T - \sigma)AW(\sigma)\,d\sigma + \int_0^t E(t - \sigma)(A + \mathrm{Id}_H)\,dW(\sigma), \quad t \in [0, T].$$

From Example 4.19 recall the identity

$$W(t) = A\int_0^t E(t - \sigma)W(\sigma)\,d\sigma + \int_0^t E(t - \sigma)\,dW(\sigma),$$

which yields

$$X(t) = W(t) + \int_0^t E(t - \sigma)A\,dW(\sigma).$$

From this, it directly follows that

$$\mathbf{E}\big[\varphi(X(T))\big] = \mathbf{E}\big[\|X(T)\|^2\big] = \sum_{j=1}^\infty \mu_j \mathbf{E}\Big[\Big|\beta_j(T) + \lambda_j \int_0^T e^{-\lambda_j(T-\sigma)}\,d\beta_j(\sigma)\Big|^2\Big]$$

$$= \sum_{j=1}^\infty \mu_j\Big(t + 2\big(1 - e^{-\lambda_j T}\big) + \frac{1}{2}\lambda_j\big(1 - e^{-2\lambda_j T}\big)\Big),$$

$$(6.19)$$

which we use for the computation of the weak error.

The spatially semidiscrete approximation $\overline{X}_h(t)$ of $X(t)$ is the solution to

$$d\overline{X}_h(t) + \big[A_h\overline{X}_h(t) + J_h f(t)\big]\,dt = J_h(A + \mathrm{Id}_H)\,dW(t), \quad \text{for } 0 \le t \le T,$$

$$\overline{X}_h(0) = 0,$$

$$(6.20)$$

where J_h denotes the finitely based interpolation operator from (6.12). Then, in terms of the orthogonal basis $(e_{h,j})_{j=1}^{N_h}$ we have the representation

$$\overline{X}_h(T) = \int_0^T E_h(T - \sigma)J_h AW(\sigma)\,d\sigma + \int_0^T E_h(T - \sigma)J_h(A + \mathrm{Id}_H)\,dW(\sigma)$$

$$= \sum_{j=1}^{N_h} \sqrt{\mu_j}\Big(\int_0^T e^{-\lambda_{h,j}(T-\sigma)}\lambda_j\beta_j(\sigma)\,d\sigma$$

$$+ \int_0^T e^{-\lambda_{h,j}(T-\sigma)}(1 + \lambda_j)\,d\beta_j(\sigma)\Big)e_{h,j}$$

$$
= \sum_{j=1}^{N_h} \sqrt{\mu_j} \Big[\frac{\lambda_j}{\lambda_{h,j}} \Big(\beta_j(T) - \int_0^T e^{-\lambda_{h,j}(T-\sigma)} \, d\beta_j(\sigma) \Big)
$$

$$
+ \int_0^T e^{-\lambda_{h,j}(T-\sigma)} (1 + \lambda_j) \, d\beta_j(\sigma) \Big] e_{h,j}
$$

$$
= \sum_{j=1}^{N_h} \sqrt{\mu_j} \Big[\frac{\lambda_j}{\lambda_{h,j}} \beta_j(T) + \Big(1 - \frac{\lambda_j}{\lambda_{h,j}} + \lambda_j \Big) \int_0^T e^{-\lambda_{h,j}(T-\sigma)} \, d\beta_j(\sigma) \Big] e_{h,j},
$$

where we again applied the identity from Example 4.19 for the third equality. From this, it is straightforward to compute

$$
\mathbf{E}[\varphi(\overline{X}_h(T))] = \mathbf{E}[\|\overline{X}_h(T)\|^2]
$$

$$
= \sum_{j=1}^{N_h} \mu_j \Big[\frac{\lambda_j^2}{\lambda_{h,j}^2} T + 2 \frac{\lambda_j}{\lambda_{h,j}} \Big(1 - \frac{\lambda_j}{\lambda_{h,j}} + \lambda_j \Big) \frac{1}{\lambda_{h,j}} \big(1 - e^{-\lambda_{h,j} T} \big)
$$

$$
+ \Big(1 - \frac{\lambda_j}{\lambda_{h,j}} + \lambda_j \Big)^2 \frac{1}{2\lambda_{h,j}} \big(1 - e^{-2\lambda_{h,j} T} \big) \Big] \|e_{h,j}\|^2.
$$

$$(6.21)$$

For the computation of the strong error we apply the same approach as in (6.15) and (6.16). For this it remains to find a computable representation of

$$
\mathbf{E}[\|\overline{X}_h(T) - J_h X(T)\|^2]
$$

$$
= \mathbf{E}[\|\overline{X}_h(T)\|^2] - 2\mathbf{E}[(\overline{X}_h(T), J_h X(T))] + \mathbf{E}[\|J_h X(T)\|^2].
$$

Since the first and third term are given by (6.21) and (6.19), respectively, we concentrate on the second term which reads

$$
\mathbf{E}[(\overline{X}_h(T), J_h X(T))]
$$

$$
= \sum_{j=1}^{N_h} \mu_j \mathbf{E}\Big[\Big(\beta_j(T) + \lambda_j \int_0^T e^{-\lambda_j(T-\sigma)} \, d\beta_j(\sigma) \Big)
$$

$$
\times \Big(\frac{\lambda_j}{\lambda_{h,j}} \beta_j(T) + \Big(1 - \frac{\lambda_j}{\lambda_{h,j}} + \lambda_j \Big) \int_0^T e^{-\lambda_{h,j}(T-\sigma)} \, d\beta_j(\sigma) \Big) \Big] \|e_{h,j}\|^2
$$

$$
= \sum_{j=1}^{N_h} \mu_j \Big[\frac{\lambda_j}{\lambda_{h,j}} T + \Big(1 - \frac{\lambda_j}{\lambda_{h,j}} + \lambda_j \Big) \frac{1}{\lambda_{h,j}} \big(1 - e^{-\lambda_{h,j} T} \big) + \frac{\lambda_j}{\lambda_{h,j}} \big(1 - e^{-\lambda_j T} \big)
$$

$$
+ \lambda_j \Big(1 - \frac{\lambda_j}{\lambda_{h,j}} + \lambda_j \Big) \frac{1}{\lambda_j + \lambda_{h,j}} \big(1 - e^{-(\lambda_j + \lambda_{h,j})T} \big) \Big] \|e_{h,j}\|^2.
$$

Table 6.2 Inhomogeneous stochastic heat equation

N_h	Strong error	EOC	Weak error	EOC
250	2.305657e−02		1.171896e−02	
500	1.134878e−02	1.022641	3.195089e−03	1.874917
1,000	5.580977e−03	1.023948	8.618200e−04	1.890398
2,000	2.743493e−03	1.024504	2.305353e−04	1.902399
4,000	1.348406e−03	1.024759	6.125544e−05	1.912078
8,000	6.626754e−04	1.024881	1.618568e−05	1.920120
16,000	3.256587e−04	1.024941	4.256578e−06	1.926952

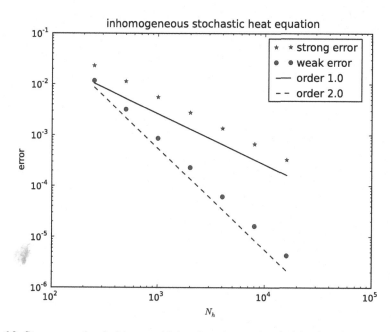

Fig. 6.2 Strong error (marked by *asterisks*) and weak error (marked by *dots*) for the semidiscretization of the inhomogeneous stochastic heat equation (values from Table 6.2)

By the help of (6.19) and (6.21) we are able to approximate the weak error in the same way as in Sect. 6.2. Since f and $g = (A + \mathrm{Id}_H)$ also satisfy the conditions of Theorem 5.12 by our choice of the eigenvalues μ_j, we expect to see that the weak error converges with order 2. As it is shown in Table 6.2 and Fig. 6.2 our numerical results are again in line with our theoretical findings.

The same is true for the order of convergence of the strong error, since $\|g\|_{L^0_{2,r}} = \|(A + \mathrm{Id}_H)\|_{L^0_{2,r}} < \infty$ for all $r \in [0, 0.025)$. Thus, our experimental order of convergence matches very well the theoretical order of 1.025.

6.4 Geometric Brownian Motion in Infinite Dimensions

Our third example is an infinite dimensional analogue of the geometric Brownian motion. That is, we consider the following stochastic evolution equation with *multiplicative noise*

$$
\begin{aligned}
\mathrm{d}X(t) + AX(t)\,\mathrm{d}t &= g(X(t))\,\mathrm{d}W(t), \quad \text{for all } t \in [0, T], \\
X(0) &= X_0,
\end{aligned}
\tag{6.22}
$$

where, as before, $-A$ is the Laplace operator with Dirichlet boundary conditions on $[0, 1]$. The initial condition X_0 is assumed to be deterministic and the linear mapping $g : H \to L_2(H, H)$ is given by

$$
g(u) = \sum_{j=1}^{\infty} (u, e_j) e_j \otimes e_j.
\tag{6.23}
$$

Indeed, g is $L_2(H, H)$-valued and it holds

$$
\|g(u)\|_{L_2(H,H)}^2 = \sum_{i=1}^{\infty} \|g(u)e_i\|^2 = \sum_{i=1}^{\infty} \|(u, e_i)e_i\|^2 = \sum_{i=1}^{\infty} (u, e_i)^2 = \|u\|^2.
$$

A similar calculation shows that $\|g(u)\|_{L_{2,r}^0} = \|A^{\frac{r}{2}} g(u) Q^{\frac{1}{2}}\|_{L_2(H,H)} \leq \|Q^{\frac{1}{2}}\|_{L(H)} \|u\|_r$ for all $r \geq 0$ and $u \in \dot{H}^r$. Hence g satisfies Assumptions 2.16 and 2.20. The mild solution to (6.22) has the form

$$
\begin{aligned}
X(t) &= E(t)X_0 + \int_0^t E(t - \sigma) g(X(\sigma))\,\mathrm{d}W(\sigma) \\
&= \sum_{j=1}^{\infty} \left(e^{-\lambda_j T}(X_0, e_j) + \left(\int_0^T E(T - \sigma) g(X(\sigma))\,\mathrm{d}W(\sigma), e_j \right) \right) e_j \\
&= \sum_{j=1}^{\infty} \left(e^{-\lambda_j T}(X_0, e_j) + \sqrt{\mu_j} \int_0^T e^{-\lambda_j (T - \sigma)}(X(\sigma), e_j)\,\mathrm{d}\beta_j(\sigma) \right) e_j.
\end{aligned}
\tag{6.24}
$$

In particular, it holds that

$$
x^j(t) := (X(t), e_j) = e^{-\lambda_j T}(X_0, e_j) + \sqrt{\mu_j} \int_0^T e^{-\lambda_j (T - \sigma)} x^j(\sigma)\,\mathrm{d}\beta_j(\sigma).
$$

Thus, for $j \in \mathbb{N}$, the real valued process $x^j(t)$ is a solution to the SODE

$$dx^j(t) + \lambda_j x^j(t)\,dt = \sqrt{\mu_j}x^j(t)\,dW(t)$$
$$x^j(0) = (X_0, e_j).$$

Consequently, $x^j(t)$ is a geometric Brownian motion and we have the representation [56, Sect. 3.5, Ex. 5.5]

$$x^j(t) = (X_0, e_j)\exp\left(-(\lambda_j + \frac{1}{2}\mu_j)t + \sqrt{\mu_j}\beta_j(t)\right), \quad \text{for } t \in [0, T],$$

and it holds

$$\mathbf{E}[|x^j(t)|^2] = (X_0, e_j)^2 \exp\left((-2\lambda_j + \mu_j)t\right), \quad \text{for } t \in [0, T].$$

From this, we derive

$$\mathbf{E}[\|X(T)\|^2] = \sum_{j=1}^{\infty}\mathbf{E}[|x^j(T)|^2] = \sum_{j=1}^{\infty}(X_0, e_j)^2 \exp\left((-2\lambda_j + \mu_j)T\right), \quad (6.25)$$

which we use in the computation of the weak and strong error. Note that (6.25) together with the boundedness of the function $[0, \infty) \ni x \mapsto x^r e^{-x}$ shows, that

$$\mathbf{P}(X(T) \in \dot{H}^r) = 1$$

for every $T > 0$ and $r \geq 0$.

As before, the spatially semidiscrete approximation $\overline{X}_h(t)$ of $X(t)$ is the solution to

$$d\overline{X}_h(t) + A_h\overline{X}_h(t)\,dt = g_h(\overline{X}_h(t))\,dW(t), \quad \text{for } 0 \leq t \leq T,$$
$$\overline{X}_h(0) = J_h X_0, \quad (6.26)$$

with A_h and J_h denoting the discrete Laplacian from Sect. 6.1 and the finitely based interpolation operator from (6.12), respectively.

Here we note, that $J_h X_0$ may also be well-defined for non-smooth initial conditions X_0. Since $J_h X_0$ is finitely based, we actually apply the interpolation operator I_h to functions, which are represented by a linear combination of finitely many eigenfunctions $(e_j)_{j=1}^{N_h}$. Since all such linear combinations are smooth, the application of the interpolation operator is well-defined for every $h \in (0, 1)$. However, in general the limit $h \to 0$ of $J_h u$ will not exist as a continuous function for non-smooth arguments u.

The discrete version $g_h\colon S_h \to L_2(Q^{\frac{1}{2}}(H), S_h)$ of g is given by

$$g_h(u_h) = \sum_{j=1}^{N_h}(u_h, e_{h,j})e_{h,j} \otimes e_j, \quad \text{for all } u_h \in S_h. \quad (6.27)$$

Then, in the same way as above, we obtain

$$X_h(t) = E_h(t) J_h X_0 + \int_0^T E_h(T - \sigma) g_h(\overline{X}_h(\sigma)) \, dW(\sigma)$$

$$= \sum_{j=1}^{N_h} \left(e^{-\lambda_{h,j} t} (X_0, e_j) + \sqrt{\mu_j} \int_0^t e^{-\lambda_{h,j}(t-\sigma)} (\overline{X}_h(\sigma), e_{h,j}) \, d\beta_j(\sigma) \right) e_{h,j}$$

Therefore, also the Fourier coefficients of $\overline{X}_h$ follow a real-valued geometric Brownian motion and it holds

$$\overline{X}_h(t) = \sum_{j=1}^{N_h} (X_0, e_j) \exp\left(-\left(\lambda_{h,j} + \frac{1}{2}\mu_j\right) t + \sqrt{\mu_j} \beta_j(t) \right) e_{h,j},$$

from which we derive

$$\mathbf{E}\big[\|\overline{X}_h(T)\|^2\big] = \sum_{j=1}^{N_h} (X_0, e_j)^2 \exp\left(\left(-2\lambda_{h,j} + \mu_j\right) T \right) \|e_{h,j}\|^2. \qquad (6.28)$$

With (6.25) and (6.28) we are in a position to compute the weak error. For the computation of the strong error we use in addition

$$\mathbf{E}\big[\|\overline{X}_h(T) - J_h X(T)\|^2\big] = \mathbf{E}\big[\|X(T)\|^2\big] - 2\mathbf{E}\big[(X(T), \overline{X}_h(T))\big] + \mathbf{E}\big[\|\overline{X}_h(T)\|^2\big]$$

and

$$\mathbf{E}\big[(\overline{X}_h(T), J_h X(T))\big] = \sum_{j=1}^{N_h} (X_0, e_j)^2 \exp\left(\left(-\lambda_{h,j} - \lambda_j + \mu_j\right) T \right) \|e_{h,j}\|^2.$$

$$(6.29)$$

For our experiment we set $(X_0, e_j) = 1$ for all $j \in \mathbb{N}$, that is, the initial condition is non-smooth and it holds $X_0 \in \dot{H}^r$ for every $r < -\frac{1}{2}$. Further, we set $T = 1$ and the eigenvalues of Q are chosen to be $\mu_j = 10\left(\frac{1}{j}\right)^{1.05}$. As already noted, these choices of X_0 and μ_j do not influence the spatial regularity of X at time $T > 0$ and we expect that the strong order of convergence is equal to 2, which is the optimal order for a finite element discretization with piecewise linear elements. Since the order of strong convergence is always a lower bound for the order of convergence of the weak error, we also expect to observe a weak order of convergence of order 2.

As the results in Table 6.3 and Fig. 6.3 show, our numerical experiment is in line with our theoretical findings.

Table 6.3 Geometric Brownian motion

N_h	Strong error	EOC	Weak error	EOC
250	9.890017e−07		1.672132e−08	
500	2.482519e−07	1.994168	4.197466e−09	1.994098
1,000	6.218791e−08	1.997099	1.051491e−09	1.997082
2,000	1.556269e−08	1.998543	2.631373e−10	1.998549
4,000	3.893392e−09	1.998992	6.581733e−11	1.999277
8,000	9.740005e−10	1.999033	1.645845e−11	1.999639
16,000	2.469542e−10	1.979679	4.115128e−12	1.999820

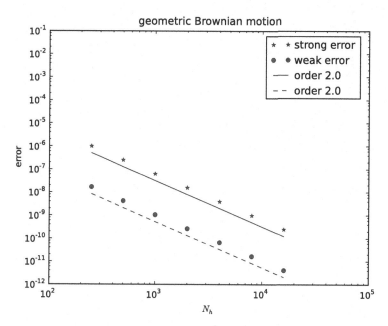

Fig. 6.3 Strong error (marked by *asterisks*) and weak error (marked by *dots*) for the semidiscretization of the geometric Brownian motion (values from Table 6.3)

6.5 Inhomogeneous Geometric Brownian Motion in Infinite Dimensions

Our final numerical experiment is concerned with an inhomogeneous variant of the geometric Brownian motion (6.22). That is, we consider the stochastic evolution equation

$$\mathrm{d}X(t) + \left[AX(t) + f\right]\mathrm{d}t = g(X(t))\,\mathrm{d}W(t), \quad \text{for all } t \in [0, T],$$
$$X(0) = 0, \tag{6.30}$$

where A and g are the same as in the previous section. In contrast to (6.22) we consider here a zero initial condition, but instead we include a constant inhomogeneity f, which coincides with our choice of $-X_0$ in Sect. 6.4, that is

$$f := -\sum_{j=1}^{\infty} e_j \in \dot{H}^{-1+r}, \quad \text{for all } r < \frac{1}{2}.$$

Further, if we follow the same arguments as in Sect. 6.4, we obtain the representation

$$X(t) = \sum_{j=1}^{\infty} \left(\int_0^t e^{-\lambda_j(t-\sigma)} d\sigma + \sqrt{\mu_j} \int_0^t e^{-\lambda_j(t-\sigma)} (X(\sigma), e_j) d\beta_j(\sigma) \right) e_j$$

$$= \sum_{j=1}^{\infty} \left(\int_0^t \exp\left(-\left(\lambda_j + \frac{1}{2}\mu_j\right)(t-\sigma) + \sqrt{\mu_j}(\beta_j(t) - \beta_j(\sigma)) \right) d\sigma \right) e_j$$

of the mild solution to 6.30, where we again used the variation of constants formula for SODEs, see [56, Chap. 3, Th. 3.1].

In this experiment the spatially semidiscrete approximation of $\overline{X}_h(t)$ of $X(t)$ is determined by the equation

$$d\overline{X}_h(t) + \left[A_h \overline{X}_h(t) + J_h f \right] dt = g_h(\overline{X}_h(t)) dW(t), \quad \text{for } 0 \le t \le T,$$
$$\overline{X}_h(0) = 0, \tag{6.31}$$

where A_h and J_h are the same as in Sect. 6.1 and $g_h \colon S_h \to L_2(Q^{\frac{1}{2}}(H), S_h)$ is given in (6.27).

In the same way as above it holds

$$\overline{X}_h(t) = \sum_{j=1}^{\infty} \left(\int_0^t e^{-\lambda_{h,j}(t-\sigma)} d\sigma + \sqrt{\mu_j} \int_0^t e^{-\lambda_j(t-\sigma)} (\overline{X}_h(\sigma), e_{h,j}) d\beta_j(\sigma) \right) e_{h,j}$$

$$= \sum_{j=1}^{\infty} \left(\int_0^t \exp\left(-\left(\lambda_{h,j} + \frac{1}{2}\mu_j\right)(t-\sigma) + \sqrt{\mu_j}(\beta_j(t) - \beta_j(\sigma)) \right) d\sigma \right) e_{h,j}.$$

As in the previous experiments, we need a computable representation of the three terms $\mathbf{E}[\|X(T)\|^2]$, $\mathbf{E}[\|\overline{X}_h(T)\|^2]$ and $\mathbf{E}[\|\overline{X}_h(T) - X(T)\|^2]$. Since all representations are derived by the same techniques, we only present all details for the first term:

$$\mathbf{E}\left[\|X(T)\|^2\right]$$

$$= \sum_{j=1}^{\infty} \mathbf{E}\left[\left| \int_0^T \exp\left(-\left(\lambda_j + \frac{1}{2}\mu_j\right)(T-\sigma) + \sqrt{\mu_j}(\beta_j(T) - \beta_j(\sigma)) \right) d\sigma \right|^2 \right]$$

$$= \sum_{j=1}^{\infty} \int_0^T \int_0^T \exp\left(-\lambda_j(T-\sigma) - \lambda_j(T-\tau)\right)$$

$$\times \mathbf{E}\left[\exp\left(-\frac{1}{2}\mu_j(2T-\sigma-\tau) + \sqrt{\mu_j}(2\beta_j(T) - \beta_j(\sigma) - \beta_j(\tau))\right)\right] d\tau \, d\sigma.$$

Note, that the integral is symmetric in the appearance of τ and σ and we split the integrals in the two equal cases $\tau < \sigma$ and $\sigma < \tau$. Then, for $\tau < \sigma$, we obtain by the independence of the increments of the Wiener process

$$\mathbf{E}\left[\exp\left(-\frac{1}{2}\mu_j(2T-\sigma-\tau) + \sqrt{\mu_j}(2\beta_j(T) - \beta_j(\sigma) - \beta_j(\tau))\right)\right]$$

$$= \mathbf{E}\left[\exp\left(-\mu_j(T-\sigma) + 2\sqrt{\mu_j}(\beta_j(T) - \beta_j(\sigma))\right)\right]$$

$$\times \mathbf{E}\left[\exp\left(-\frac{1}{2}\mu_j(\sigma-\tau) + \sqrt{\mu_j}(\beta_j(\sigma) - \beta_j(\tau))\right)\right]$$

$$= \exp\left(\mu_j(T-\sigma)\right),$$

where in the last step we applied the formula for the first and second moment of a geometric Brownian motion from [56, Sect. 3.5, Ex. 5.5]. Hence,

$$\mathbf{E}\left[\|X(T)\|^2\right]$$

$$= 2\sum_{j=1}^{\infty} \int_0^T \int_0^{\sigma} \exp\left(-2\lambda_j(T-\sigma) - \lambda_j(\sigma-\tau) + \mu_j(T-\sigma)\right) d\tau \, d\sigma$$

$$= 2\sum_{j=1}^{\infty} \int_0^T \exp\left((-2\lambda_j + \mu_j)(T-\sigma)\right) \frac{1}{\lambda_j}\left(1 - \exp(-\lambda_j\sigma)\right) d\sigma.$$

Since for $a \in \mathbb{R}, b > 0$,

$$\int_0^T \exp(a(T-\sigma))\frac{1}{b}(1 - \exp(-b\sigma)) \, d\sigma = \frac{ae^{-bT} - be^{aT} - a - b}{a^2b + ab^2},$$

we obtain

$$\mathbf{E}\left[\|X(T)\|^2\right] = 2\sum_{j=1}^{\infty} \frac{(-2\lambda_j + \mu_j)e^{-\lambda_j T} - \lambda_j e^{(-2\lambda_j + \mu_j)T} + \lambda_j - \mu_j}{(-2\lambda_j + \mu_j)^2\lambda_j + (-2\lambda_j + \mu_j)\lambda_j^2}$$

$$= 2\sum_{j=1}^{\infty} \frac{(-2\lambda_j + \mu_j)e^{-\lambda_j T} - \lambda_j e^{(-2\lambda_j + \mu_j)T} + \lambda_j - \mu_j}{2\lambda_j^3 - 3\lambda_j^2\mu_j + \mu_j^2\lambda_j}.$$

$$(6.32)$$

Table 6.4 Inhomogeneous geometric Brownian motion

N_h	Strong error	EOC	Weak error	EOC
250	2.999918e−05		1.716336e−06	
500	1.058052e−05	1.503513	4.313909e−07	1.992264
1,000	3.736100e−06	1.501805	1.081374e−07	1.996131
2,000	1.320027e−06	1.500966	2.707063e−08	1.998065
4,000	4.670021e−07	1.499066	6.772198e−09	1.999032
8,000	1.663555e−07	1.489159	1.693618e−09	1.999516
16,000	6.369910e−08	1.384925	4.234756e−10	1.999757

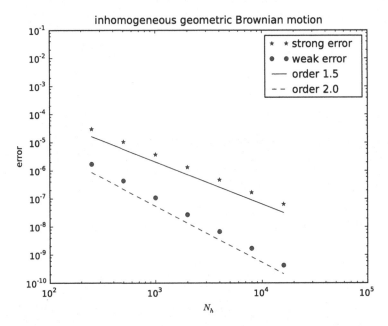

Fig. 6.4 Strong error (marked by *asterisks*) and weak error (marked by *dots*) for the semidiscretization of the inhomogeneous geometric Brownian motion (values from Table 6.4)

In the same way we get

$$\mathbf{E}\big[\|\overline{X}_h(T)\|^2\big] = 2 \sum_{j=1}^{\infty} \frac{(-2\lambda_{h,j} + \mu_j)e^{-\lambda_{h,j}T} - \lambda_{h,j}e^{(-2\lambda_{h,j}+\mu_j)T} + \lambda_{h,j} - \mu_j}{2\lambda_{h,j}^3 - 3\lambda_{h,j}^2\mu_j + \mu_j^2\lambda_{h,j}}.$$

(6.33)

For the computation we again set $T = 1$ and the eigenvalues of Q are chosen to be $\mu_j = 10\left(\frac{1}{j}\right)^{1.05}$. By our above choice of f, the spatial regularity theorem then yields that the mild solution X takes values in $\dot{H}^{1+r}$ for every $r < \frac{1}{2}$. Hence, we also expect to observe a strong order of convergence of 1.5. On the other hand, our result on the weak error is not applicable since we have an equation with multiplicative noise. Nevertheless, we expect to see a weak order of convergence somewhere in the interval $[1.5, 2]$, since the strong error forms a lower bound and the optimal order of convergence for the linear finite element approximation is 2.

The results of our experiment are shown in Table 6.4 and Fig. 6.4. In fact, the strong order behaves a little bit worse then expected, especially for large values of N_h. On the other hand, the weak error behaves very well and converges with the optimal order of 2.

The closure of the gap between the theory and the numerical experiment for multiplicative noise is subject to future research.

Appendix A
Some Useful Variations of Gronwall's Lemma

In (numerical) analysis of differential equations Gronwall's Lemma plays an important role. The original version is due to T.H. Grönwall [31], but there exist a huge number of variations. In this chapter we collect some useful versions and provide those proofs which are not easily found elsewhere in the literature.

The first result is a standard integral version of Gronwall's Lemma, which together with a proof can be found in, for example, [56, Chap. 1,Th. 8.1].

Lemma A.1. *Let* $T > 0$ *and* $c \geq 0$. *Let* $\varphi, v : [0, T] \rightarrow \mathbb{R}$ *be continuous and nonnegative functions. If*

$$\varphi(t) \leq c + \int_0^t v(\sigma)\varphi(\sigma)\,d\sigma \quad \text{for all } t \in [0, T],$$

then

$$\varphi(t) \leq c \exp\left(\int_0^t v(\sigma)\,d\sigma \right) \quad \text{for all } t \in [0, T].$$

The next variation of Gronwall's Lemma is a generalization with weak singularities. A reference for this version is, for example, [36, Lemma 7.1.1], while the presented proof follows [23].

Lemma A.2. *Let* $T > 0$ *and* $C_1, C_2 \geq 0$ *and let* $\varphi : [0, T] \rightarrow \mathbb{R}$ *be a nonnegative and continuous function. Let* $\beta > 0$. *If we have*

$$\varphi(t) \leq C_1 + C_2 \int_0^t (t - \sigma)^{-1+\beta}\varphi(\sigma)\,d\sigma \quad \text{for all } t \in (0, T], \tag{A.1}$$

then there exists a constant $C = C(C_2, T, \beta)$ *such that*

$$\varphi(t) \leq C C_1, \quad \text{for all } t \in (0, T].$$

R. Kruse, *Strong and Weak Approximation of Semilinear Stochastic Evolution Equations,* 155
Lecture Notes in Mathematics 2093, DOI 10.1007/978-3-319-02231-4,
© Springer International Publishing Switzerland 2014

Proof. For the proof we recall the following identity

$$\int_0^t (t-\sigma)^{-1+\alpha}\sigma^{-1+\beta}\,d\sigma = B(\alpha,\beta)t^{-1+\alpha+\beta}, \tag{A.2}$$

where $B(x,y)$ denotes the beta function.

Choose the smallest $n = n(\beta) \in \mathbb{N}$ such that $-1 + n\beta \geq 0$ and iterate the inequality (A.1) $n-1$ times, then by applying (A.2) we obtain

$$\varphi(t) \leq D_1 C_1 + C_2 \int_0^t (t-\sigma)^{-1+n\beta}\varphi(\sigma)\,d\sigma$$

$$\leq D_1 C_1 + D_2 T^{-1+n\beta} \int_0^t \varphi(\sigma)\,d\sigma,$$

with constants $D_1 = D_1(C_2, T, \beta)$ and $D_2 = D_2(C_2, \beta)$. Now Lemma A.1 yields the desired results. $\qquad\qquad\qquad\qquad\qquad\qquad\qquad\qquad\qquad\qquad\qquad\qquad\square$

While the last two lemmas yield estimates for continuous functions, there also exist discrete analogues. The next lemma is a slightly generalized version of [30, 2.2. (9)] (see J.M. Holte[1] for a more recent presentation).

Lemma A.3. *Let $c \geq 0$ and $(\varphi_j)_{j\geq 1}$ and $(v_j)_{j\geq 1}$ be nonnegative sequences. If*

$$\varphi_j \leq c + \sum_{i=1}^{j-1} v_i\varphi_i \quad \text{for } j \geq 1,$$

then

$$\varphi_j \leq c \prod_{i=1}^{j-1}(1+v_i) \leq c\exp\left(\sum_{i=1}^{j-1} v_i\right) \quad \text{for } j \geq 1.$$

We also have a discrete version of Lemma A.2. Here we follow the proof of [23, Lemma 7.1].

Lemma A.4. *For $T > 0$ and $k > 0$ consider $t_j = jk$ with $j = 1,\ldots,N_k$ such that $N_k k \leq T < (N_k + 1)k$. Let $C_1, C_2 \geq 0$ and let $(\varphi_j)_{j=1,\ldots,N_k}$ be a nonnegative sequence.*

If for $\beta \in (0,1]$ we have

$$\varphi_j \leq C_1 + C_2 k \sum_{i=1}^{j-1} t_{j-i}^{-1+\beta}\varphi_i \quad \text{for all } j = 1,\ldots,N_k, \tag{A.3}$$

---

[1]http://homepages.gac.edu/~holte/publications/gronwallTALK.pdf

then there exists a constant $C = C(C_2, T, \beta)$ such that

$$\varphi_j \leq C C_1 \quad \text{for all } j = 1, \ldots, N_k.$$

In particular, the constant C does not depend on k.

Proof. By using (A.2) we get for $\alpha, \beta \in (0, 1]$

$$k \sum_{i=0}^{j-1} t_{j-i}^{-1+\alpha} t_{i+1}^{-1+\beta} \leq \sum_{i=0}^{j-1} \int_{t_i}^{t_{i+1}} (t_j - \sigma)^{-1+\alpha} \sigma^{-1+\beta} \, d\sigma$$

$$\leq \int_0^{t_j} (t_j - \sigma)^{-1+\alpha} \sigma^{-1+\beta} \, d\sigma = B(\alpha, \beta) t_j^{-1+\alpha+\beta}. \tag{A.4}$$

As in the proof of Lemma A.2 we choose the smallest $n = n(\beta) \in \mathbb{N}$ such that $-1 + n\beta \geq 0$ and iterate the inequality (A.3) $n - 1$ times. Then by applying (A.4) we obtain

$$\varphi_j \leq D_1 C_1 + D_2 k \sum_{i=1}^{j-n} t_{j-i}^{-1+n\beta} \varphi_i$$

$$\leq D_1 C_1 + D_2 T^{-1+n\beta} k \sum_{i=1}^{j-1} \varphi_i,$$

with constants $D_1 = D_1(C_2, T, \beta)$ and $D_2 = D_2(C_2, \beta)$. Now the desired result follows by an application of Lemma A.3. □

Appendix B
Results on Semigroups and Their Infinitesimal Generators

The first section of this chapter provides a short review of the theory of strongly continuous semigroups on Banach spaces. The content is primarily based on [60, 69]. The second section deals with semigroups on Hilbert spaces, whose infinitesimal generators are self-adjoint and have compact inverses. In this case we define fractional powers $(-A)^r$, $r \in (0, \infty)$, of the generator and introduce a characterization of the dual space of the domain $\text{dom}((-A)^r)$.

B.1 Strongly Continuous Semigroups of Bounded Operators

In this section we consider a Banach space B.

Definition B.1. A family $(E(t))_{t \in [0,\infty)}$ of bounded linear operators from B into B is called a *strongly continuous semigroup* (or a C_0-*semigroup*) if:

(i) $E(0) = \text{Id}_B$,
(ii) $E(t + s) = E(t)E(s)$ for all $t, s \geq 0$,
(iii) $\lim_{t \searrow 0} E(t)x = x$ for all $x \in B$.

Strongly continuous semigroups enjoy the following properties.

Lemma B.2. *Let $(E(t))_{t \in [0,\infty)}$ be a C_0-semigroup on a Banach space B. There exist constants $c \geq 0$ and $M \geq 1$ such that*

$$\| E(t) \| \leq M e^{ct}$$

for all $t \in [0, \infty)$.

For the proof we refer to [60, Chap. 1.2, Th. 2.2]. If $c = 0$ the semigroup is *uniformly bounded*. In addition, if we also have $M = 1$ then we call $(E(t))_{t \in [0,\infty)}$ a *semigroup of contractions*.

R. Kruse, *Strong and Weak Approximation of Semilinear Stochastic Evolution Equations*, Lecture Notes in Mathematics 2093, DOI 10.1007/978-3-319-02231-4, © Springer International Publishing Switzerland 2014

Lemma B.3. *Let $(E(t))_{t\in[0,\infty)}$ be a C_0-semigroup on a Banach space B. Then the mapping*

$$[0, \infty) \times B \to B, \quad (t, x) \mapsto E(t)x$$

is continuous. In particular, for every $x \in B$ the mapping $t \mapsto E(t)x$ is uniformly continuous on compact subintervals of $[0, \infty)$.

A proof of this lemma is found in [69, Lem. VII.4.3] or [60, Chap. 1.2, Cor. 2.3].

The next definition assigns a linear operator to a semigroup.

Definition B.4. Consider a C_0-semigroup $(E(t))_{t\in[0,\infty)}$ on a Banach space B. The linear operator A defined by

$$Ax = \lim_{h\searrow 0} \frac{E(h)x - x}{h}$$

with domain

$$\mathrm{dom}(A) = \left\{ x \in B : \lim_{h\searrow 0} \frac{E(h)x - x}{h} \text{ exists in } B \right\}$$

is called the *infinitesimal generator* of the semigroup $(E(t))_{t\in[0,\infty)}$.

Lemma B.5. *Let $(E(t))_{t\in[0,\infty)}$ be a C_0-semigroup on a Banach space B with infinitesimal generator A. Then we have the following properties:*

(i) For all $x \in B$ it holds that

$$\lim_{h\to 0} \frac{1}{h} \int_t^{t+h} E(\sigma)x \, \mathrm{d}\sigma = E(t)x, \quad \text{for all } t \in [0, \infty).$$

(ii) For all $x \in B$ we have $\int_0^t E(\sigma)x \, \mathrm{d}\sigma \in \mathrm{dom}(A)$ and

$$E(t)x - x = A\left(\int_0^t E(\sigma)x \, \mathrm{d}\sigma \right), \quad \text{for all } t \in [0, \infty).$$

(iii) For all $x \in \mathrm{dom}(A), t \in [0, \infty)$ we have $E(t)x \in \mathrm{dom}(A)$ and

$$\frac{\mathrm{d}}{\mathrm{d}t} E(t)x = AE(t)x = E(t)Ax.$$

(iv) For all $x \in \mathrm{dom}(A)$ and $s, t \in [0, \infty), s < t$, it holds that

$$E(t)x - E(s)x = \int_s^t AE(\sigma)x \, \mathrm{d}\sigma = \int_s^t E(\sigma)Ax \, \mathrm{d}\sigma.$$

This lemma coincides with [60, Chap. 1.2, Th. 2.4]. As the next result from [60, Chap. 1.2, Th. 2.6] shows a semigroup is uniquely determined by its infinitesimal generator.

Lemma B.6. *Two C_0-semigroups with the same infinitesimal generator A coincide.*

The following theorem gives a characterization of the infinitesimal generator of a C_0-semigroup of contractions.

Theorem B.7 (Hille–Yosida). *A linear, possibly unbounded operator A: $\mathrm{dom}(A)\subset B \rightarrow B$ is the infinitesimal generator of a C_0-semigroup of contractions $(E(t))_{t\in[0,\infty)}$ if and only if:*

(i) A is closed and $\mathrm{dom}(A)$ is dense in B.
(ii) The resolvent set $\rho(A)$ of A contains the positive real line and

$$\|R(\lambda, A)\| = \|(\lambda I - A)^{-1}\| \leq \frac{1}{\lambda}, \quad \text{for all } \lambda > 0.$$

A proof is given in [60, Chap. 1.3, Th. 3.1] and [69, Th. VII.4.11]. A version of this theorem, which gives a corresponding characterization of the infinitesimal generator of general C_0-semigroups, is found in [69, Th. VII.4.13].

B.2   Fractional Powers of A and the Spaces $\dot{H}^s$

This section deals with semigroups on a separable Hilbert space H with inner product $(\cdot, \cdot)$ and norm $\|\cdot\|$. We consider a densely defined, linear, self-adjoint and positive definite operator A: $\mathrm{dom}(A) \subset H \rightarrow H$, which is not necessarily bounded but with compact inverse. Under these conditions Theorem B.7 yields that $-A$ is the infinitesimal generator of a C_0-semigroup of contractions $(E(t))_{t\in[0,\infty)}$ on H.

By applying the spectral theorem for linear compact and self-adjoint operators [69, Th. VI.3.2] to A^{-1} we obtain the existence of an increasing sequence of real numbers $(\lambda_n)_{n\geq 1}$ and an orthonormal basis of eigenvectors $(e_n)_{n\geq 1}$ in H such that $Ae_n = \lambda_n e_n, n \in \mathbb{N}$, and

$$0 < \lambda_1 \leq \lambda_2 \leq \ldots \leq \lambda_n(\rightarrow \infty).$$

We have the following characterization of the domain of A

$$\mathrm{dom}(A) = \left\{x \in H : \sum_{n=1}^{\infty}\lambda_n^2(x, e_n)^2 < \infty\right\}.$$

In fact, from [60, Chap. 2.5, Th. 5.2] it follows that $(E(t))_{t\in[0,\infty)}$ is an analytic semigroup.

The above conditions on A are more restrictive as in [60, Chap. 2.6] but they allow us to define fractional powers of A in a much simpler way (see also [57, Ex. 6.1.2, Ex. 6.1.7]). For any $r \geq 0$ let the operator $A^{\frac{r}{2}}: \mathrm{dom}(A^{\frac{r}{2}}) \subset H \to H$ be given by

$$A^{\frac{r}{2}}x = \sum_{n=1}^{\infty} \lambda_n^{\frac{r}{2}}(x, e_n)e_n \tag{B.1}$$

for all

$$x \in \mathrm{dom}(A^{\frac{r}{2}}) = \left\{ x \in H : \|x\|_r^2 := \sum_{n=1}^{\infty} \lambda_n^r(x, e_n)^2 < \infty \right\}.$$

By setting $\dot{H}^r := \mathrm{dom}(A^{\frac{r}{2}})$ and $(\cdot, \cdot)_r := (A^{\frac{r}{2}}\cdot, A^{\frac{r}{2}}\cdot)$ we obtain a separable Hilbert space $(\dot{H}^r, (\cdot, \cdot)_r, \|\cdot\|_r)$ for every $r > 0$.

The next result gives a characterization of the dual space $(\dot{H}^r)'$ with $r > 0$. For this we consider the set

$$\dot{H}^{-r} := \left\{ x = \sum_{n=1}^{\infty} x_n e_n : x_n \in \mathbb{R}, n = 1, 2, \ldots, \right.$$

$$\left. \text{such that } \|x\|_{-r}^2 := \sum_{n=1}^{\infty} \lambda_n^{-r} x_n^2 < \infty \right\}$$

and, analogously to (B.1), we define the fractional power of A for negative exponents by

$$A^{-\frac{r}{2}}x = \sum_{n=1}^{\infty} \lambda_n^{-\frac{r}{2}} x_n e_n$$

for all $x = \sum_{n=1}^{\infty} x_n e_n \in \dot{H}^{-r}$. It always holds that $H \subset \dot{H}^{-r}$ but in general we have $H \neq \dot{H}^r$ for every $r > 0$. It follows that $\dot{H}^{-r}$ is the largest set such that $A^{-\frac{r}{2}}$ maps into H. In this sense $\dot{H}^{-r} = \mathrm{dom}(A^{-\frac{r}{2}})$.

As above we endow $\dot{H}^{-r}$ with the inner product $(\cdot, \cdot)_{-r} := (A^{-\frac{r}{2}}\cdot, A^{-\frac{r}{2}}\cdot)$ and the norm $\|\cdot\|_{-r} = \|A^{-\frac{r}{2}}\cdot\|$.

Theorem B.8. *For $r > 0$ the dual space $(\dot{H}^r)'$ is isometrically isomorphic to $\dot{H}^{-r}$. In particular, $\dot{H}^{-r}$ is a separable Hilbert space.*

Proof. We follow the lines of the proof of [69, Th. II.2.3] together with some suitable generalizations.

Let us define a linear operator $T: \dot{H}^{-r} \to (\dot{H}^r)'$ by

$$(Tx)(y) = \sum_{n=1}^{\infty} x_n(y, e_n)$$

for $x = \sum_{n=1}^{\infty} x_n e_n \in \dot{H}^{-r}$ and $y = \sum_{n=1}^{\infty} (y, e_n) e_n \in \dot{H}^r$. Clearly, $T: \dot{H}^{-r} \to (\dot{H}^r)'$ and $Tx: \dot{H}^r \to \mathbb{R}$ are both linear mappings for every $x \in \dot{H}^{-r}$. Further,

$$|(Tx)(y)| = \left| \sum_{n=1}^{\infty} x_n (y, e_n) \right| = \left| \sum_{n=1}^{\infty} \lambda_n^{-\frac{r}{2}} x_n \lambda_n^{\frac{r}{2}} (y, e_n) \right|$$

$$\leq \left(\sum_{n=1}^{\infty} \lambda_n^{-r} x_n^2 \right)^{\frac{1}{2}} \left(\sum_{n=1}^{\infty} \lambda_n^r (y, e_n)^2 \right)^{\frac{1}{2}} = \|x\|_{-r} \|y\|_r .$$

Consequently,

$$\sup_{\substack{y \in \dot{H}^r \\ \|y\|_r = 1}} |(Tx)(y)| \leq \|x\|_{-r},$$

which shows that T indeed maps into $(\dot{H}^r)'$.

Next, we prove that T is one-to-one. For this we consider $x \in \dot{H}^{-r}$ such that $Tx = 0 \in (\dot{H}^r)'$. Then we obtain

$$0 = (Tx)(e_n) = x_n \quad \text{for every } n = 1, 2, \ldots.$$

Hence, $x = 0 \in \dot{H}^{-r}$.

It remains to show that T is onto and isometric. For this we consider an arbitrary element $z \in (\dot{H}^r)'$. We set $x_n := z(e_n), n = 1, 2, \ldots$, and prove that

$$x := \sum_{n=1}^{\infty} x_n e_n \in \dot{H}^{-r}, \quad \|x\|_{-r} \leq \sup_{\substack{y \in \dot{H}^r \\ \|y\|_r = 1}} |z(y)| \tag{B.2}$$

and $Tx = z$. Let us define

$$y_n := \lambda_n^{-r} x_n \quad \text{for } n = 1, 2, \ldots.$$

Then, it holds for every $N \in \mathbb{N}$

$$0 \leq \sum_{n=1}^{N} \lambda_n^{-r} x_n^2 = \sum_{n=1}^{N} \lambda_n^r y_n^2$$

$$= \sum_{n=1}^{N} y_n x_n = \sum_{n=1}^{N} y_n z(e_n) = z \left(\sum_{n=1}^{N} y_n e_n \right)$$

$$\leq \sup_{\substack{y \in \dot{H}^r \\ \|y\|_r = 1}} |z(y)| \left\| \sum_{n=1}^{N} y_n e_n \right\|_r = \sup_{\substack{y \in \dot{H}^r \\ \|y\|_r = 1}} |z(y)| \left(\sum_{n=1}^{N} \lambda_n^r y_n^2 \right)^{\frac{1}{2}} .$$

Therefore,

$$\left(\sum_{n=1}^{N} \lambda_n^{-r} x_n^2\right)^{\frac{1}{2}} = \left(\sum_{n=1}^{N} \lambda_n^{r} y_n^2\right)^{\frac{1}{2}} \leq \sup_{\substack{y \in \dot{H}^r \\ \|y\|_r = 1}} |z(y)|$$

for every $N \in \mathbb{N}$. By taking the limit $N \to \infty$ we obtain (B.2).

In addition, it holds

$$(Tx)(e_n) = x_n = z(e_n) \quad \text{for all } n = 1, 2, \ldots$$

and both linear mappings Tx and z coincide for every $N \in \mathbb{N}$ on span$\{e_n : n = 1, \ldots, N\}$. Since both mappings are continuous in $\dot{H}^r$ they also coincide on the closure of span$\{e_n : n = 1, 2, \ldots\}$ with respect to the norm $\| \cdot \|_r$. But this closure is equal to $\dot{H}^r$, which completes the proof.   □

Next, it is worth noting that the spectral structure of A carries over to the semigroup $(E(t))_{t \in [0,T]}$. In fact, by setting

$$\tilde{E}(t)x := \sum_{n=1}^{\infty} e^{-\lambda_n t}(x, e_n)e_n \tag{B.3}$$

for all $x \in H$ and $t \in [0, \infty)$, we obtain a C_0-semigroup on H. Consider $x \in H$ such that

$$y := \lim_{h \searrow 0} \frac{\tilde{E}(h)x - x}{h} \quad \text{exists in } H. \tag{B.4}$$

Then, we get by Parseval's identity

$$0 = \lim_{h \searrow 0} \left\| \frac{\tilde{E}(h)x - x}{h} - y \right\|^2 = \lim_{h \searrow 0} \sum_{n=1}^{\infty} \left(\frac{e^{-\lambda_n h} - 1}{h}(x, e_n) - (y, e_n) \right)^2,$$

which yields

$$(y, e_n) = \lim_{h \searrow 0} \frac{e^{-\lambda_n h} - 1}{h}(x, e_n) = -\lambda_n(x, e_n).$$

Since $y \in H$ it follows that $x \in \text{dom}(A) = \dot{H}^2$ and $y = -Ax$.

That the limes in (B.4) exists for all $x \in \text{dom}(A)$ follows in a similar way by applying Lebesgue's dominated convergence theorem. Therefore, the infinitesimal generator of $\tilde{E}$ coincides with $-A$ and from Lemma B.6 we obtain that (B.3) is indeed a spectral representation of E.

The next lemma gives some very useful norm estimates of $(E(t))_{t\in[0,\infty)}$. Since we make use of them very frequently, we present a proof, but only under the above conditions on A. The estimates *(i)*, *(ii)* and *(iv)* are also valid for analytic semigroups in general. For this we refer to [60, Chap. 2.6, Th. 6.13] and [57, Th. 6.1.8].

Lemma B.9. *Under the above conditions on the infinitesimal generator $-A$ of the semigroup $(E(t))_{t\in[0,\infty)}$ the following properties hold true:*

(i) For any $\mu \geq 0$ it holds that

$$A^\mu E(t)x = E(t)A^\mu x \quad \text{for all } x \in \dot{H}^{2\mu}$$

and there exists a constant $C = C(\mu)$ such that

$$\|A^\mu E(t)\| \leq Ct^{-\mu} \quad \text{for } t > 0.$$

(ii) For any $0 \leq \nu \leq 1$ there exists a constant $C = C(\nu)$ such that

$$\|A^{-\nu}(E(t) - \mathrm{Id}_H)\| \leq Ct^\nu \text{ for } t \geq 0.$$

(iii) For any $0 \leq \rho \leq 1$ there exists a constant $C = C(\rho)$ such that

$$\int_{\tau_1}^{\tau_2} \|A^{\frac{\rho}{2}} E(\tau_2 - \sigma)x\|^2 \, d\sigma \leq C(\tau_2 - \tau_1)^{1-\rho} \|x\|^2 \text{ for all } x \in H, 0 \leq \tau_1 < \tau_2.$$

(iv) For any $0 \leq \rho \leq 1$ there exists a constant $C = C(\rho)$ such that

$$\left\| A^\rho \int_{\tau_1}^{\tau_2} E(\tau_2 - \sigma)x \, d\sigma \right\| \leq C(\tau_2 - \tau_1)^{1-\rho} \|x\| \text{ for all } x \in H, 0 \leq \tau_1 < \tau_2.$$

Proof. The first part of *(i)* follows directly from the spectral representations (B.1) and (B.3).

In order to prove the second part of *(i)* we make use of the fact that the function $x \mapsto x^\mu e^{-x}$ is bounded for $x \in [0, \infty)$. Consequently, for a constant $C = C(\mu) > 0$ we have

$$\|A^\mu E(t)\| = \sup_{n\geq 1} |\lambda_n^\mu e^{-t\lambda_n}| \leq Ct^{-\mu}.$$

For *(ii)* we apply the fact that the function $x \mapsto x^{-\nu}(1 - e^{-x})$ is bounded for $x \in [0, \infty)$ and $\nu \in [0, 1]$. Hence, for a constant $C > 0$ which only depends on $\nu \in [0, 1]$

$$\|A^{-\nu}(E(t) - \mathrm{Id}_H)\| = \sup_{n\geq 1} \left| \frac{1 - e^{-t\lambda_n}}{\lambda_n^\nu} \right| \leq Ct^\nu.$$

For the proof of *(iii)* we use the expansion of $x \in H$ in terms of the eigenbasis $(e_n)_{n\geq 1}$ of the operator A. By Parseval's identity we get

$$\int_{\tau_1}^{\tau_2} \left\| A^{\frac{\rho}{2}} E(\tau_2 - \sigma)x \right\|^2 d\sigma = \int_{\tau_1}^{\tau_2} \left\| \sum_{n=1}^{\infty} A^{\frac{\rho}{2}} E(\tau_2 - \sigma)(x, e_n)e_n \right\|^2 d\sigma$$

$$= \sum_{n=1}^{\infty} \int_{\tau_1}^{\tau_2} (x, e_n)^2 \lambda_n^\rho e^{-2\lambda_n(\tau_2-\sigma)} d\sigma$$

$$= \frac{1}{2} \sum_{n=1}^{\infty} (x, e_n)^2 \lambda_n^{\rho-1} \left(1 - e^{-2\lambda_n(\tau_2-\tau_1)}\right).$$

Again, by the boundedness of the function $x \mapsto x^{\rho-1}(1 - e^{-x})$ for $x \in [0, \infty)$ and $\rho \in [0, 1]$ there exists a constant $C = C(\rho) > 0$ such that

$$\int_{\tau_1}^{\tau_2} \left\| A^{\frac{\rho}{2}} E(\tau_2 - \sigma)x \right\|^2 d\sigma \leq C(\rho)(\tau_2 - \tau_1)^{1-\rho} \sum_{n=1}^{\infty} (x, e_n)^2,$$

which completes the proof of *(iii)*.

The following proof of *(iv)* also works for analytic semigroups in general. By Lemma B.5*(ii)* we first notice that

$$\left\| A^\rho \int_{\tau_1}^{\tau_2} E(\tau_2 - \sigma)x \, d\sigma \right\| = \left\| A^{\rho-1} A \int_0^{\tau_2-\tau_1} E(\sigma)x \, d\sigma \right\|$$

$$= \left\| A^{\rho-1}\left(E(\tau_2 - \tau_1) - I\right)x \right\|$$

Then, *(iv)* follows from *(ii)*. □

The following result is concerned with the continuity of the semigroup in the border case $\rho = 1$ of Lemma B.9*(iii)* and *(iv)*.

Lemma B.10. *Let $0 \leq \tau_1 < \tau_2$. Then we have:*

(i)

$$\lim_{\tau_2-\tau_1 \to 0} \int_{\tau_1}^{\tau_2} \left\| A^{\frac{1}{2}} E(\tau_2 - \sigma)x \right\|^2 d\sigma = 0 \quad \text{for all } x \in H,$$

(ii)

$$\lim_{\tau_2-\tau_1 \to 0} \left\| A \int_{\tau_1}^{\tau_2} E(\tau_2 - \sigma)x \, d\sigma \right\| = 0 \quad \text{for all } x \in H.$$

Proof. As in the proof of Lemma B.9 we use the orthogonal expansion of $x \in H$ with respect to the eigenbasis $(e_n)_{n \geq 1}$ of the operator A. Thus, for *(i)* we get, as in the proof of Lemma B.9*(iii)*,

$$\int_{\tau_1}^{\tau_2} \left\| A^{\frac{1}{2}} E(\tau_2 - \sigma)x \right\|^2 d\sigma = \frac{1}{2} \sum_{n=1}^{\infty} (x, e_n)^2 \left(1 - e^{-2\lambda_n(\tau_2 - \tau_1)}\right).$$

We apply Lebesgue's dominated convergence theorem. Note that the sum is dominated by $\frac{1}{2}\|x\|^2$ for all $\tau_2 - \tau_1 \geq 0$. Moreover, for every $n \geq 1$ we have

$$\lim_{\tau_2 - \tau_1 \to 0} \left(1 - e^{-2\lambda_n(\tau_2 - \tau_1)}\right)(x, e_n)^2 = 0.$$

Hence, Lebesgue's theorem gives us *(i)*.

The case *(ii)* is actually true for all strongly continuous semigroups, since by Lemma B.5*(ii)* and Lemma B.3 it holds

$$\lim_{\tau_2 - \tau_1 \to 0} \left\| A \int_{\tau_1}^{\tau_2} E(\tau_2 - \sigma)x \, d\sigma \right\| = \lim_{\tau_2 - \tau_1 \to 0} \left\| E(\tau_2 - \tau_1)x - x \right\| = 0.$$

The proof is complete. $\qquad\square$

We close this section with an extension of the linear operators $(E(t))_{t \in [0,\infty)}$ to the spaces $\dot{H}^r$ with $r < 0$. In the same way as in (B.3) we define

$$E^r(t) := \sum_{n=1}^{\infty} e^{-\lambda_n t} x_n e_n \qquad (B.5)$$

for all $t \in [0, \infty)$ and $x = \sum_{n=1}^{\infty} x_n e_n \in \dot{H}^r$ with $r < 0$. As above, the family of linear operators $(E^r(t))_{t \in [0,\infty)}$ is a strongly continuous semigroup on $\dot{H}^r$ and for all $x \in H$ it holds that $E^r(t)x = E(t)x$. More precisely, it holds that $E(t)$ and $E^r(t)$ are similar, that is

$$E^r(t)x = A^{\frac{r}{2}} E(t) A^{-\frac{r}{2}},$$

for all $t > 0$, which implies that $E(t)$ and $E^r(t)$ have the same spectrum [24, Chap. II, Cor. 5.3]. Further, as in [24, Chap. II, Cor. 5.5] one can show that the infinitesimal generator of $(E^r(t))_{t \in [0,\infty)}$ is the unique continuous extension of A to an isometry from $\dot{H}^{r+2}$ to $\dot{H}^r$.

In most occasions we drop the index r in the notation of $E^r(t)$ and also write $E(t)$ for the extended semigroup.

Appendix C
A Generalized Version of Lebesgue's Theorem

Lebesgue's dominated convergence theorem is an important and well-known tool in measure theory and probability. However, the standard formulation of the theorem turns out to be too restrictive for some proofs in this book and we rely on the following generalized version which is due to H.W. Alt [1] and [2, 1.23].

Theorem C.1. *Let $(S, \mathcal{B}, \mu)$ denote a measure space and Y a Banach space with norm $|\cdot|$. Consider Borel measurable mappings $f, f_n \colon S \to Y$, $n = 1, 2, \ldots$, and mappings $g, g_n \in L^1(S; \mathbb{R})$, $n = 1, 2, \ldots$, such that $g_n \to g$ as $n \to \infty$ with respect to the norm in $L^1(S; \mathbb{R})$. If it holds that*

$$|f_n| \le |g_n| \qquad \mu\text{-almost everywhere for all } n \in \mathbb{N} \text{ and}$$
$$f_n \to f \qquad \mu\text{-almost everywhere for } n \to \infty,$$

then $f, f_n \in L^1(S; Y)$ and $f_n \to f$ in $L^1(S; Y)$ as $n \to \infty$.
In particular, it follows that

$$\lim_{n \to \infty} \int_S f_n \, d\mu = \int_S \lim_{n \to \infty} f_n \, d\mu.$$

For the proof we refer to [1] and [2, 1.23].

R. Kruse, *Strong and Weak Approximation of Semilinear Stochastic Evolution Equations*, 169
Lecture Notes in Mathematics 2093, DOI 10.1007/978-3-319-02231-4,
© Springer International Publishing Switzerland 2014

References

1. H.W. Alt, Analysis III. Script University of Bonn. Lecture winter term (2001/2002), www. iam.uni-bonn.de/~alt/ws2001/EN/analysis3-hyp.html
2. H.W. Alt, *Lineare Funktionalanalysis*, 5th revised edn. (Springer, Berlin, 2006)
3. A. Barth, A. Lang, Milstein approximation for advection-diffusion equations driven by multiplicative noncontinuous martingale noises. Appl. Math. Optim. **66**(3), 387–413 (2012)
4. A. Barth, A. Lang, Multilevel Monte Carlo method with applications to stochastic partial differential equations. J. Comput. Math. **89**(18), 2479–2498 (2012)
5. A. Barth, A. Lang, Ch. Schwab, Multilevel Monte Carlo method for parabolic stochastic partial differential equations. BIT Numer. Math. **53**(1), 3–27 (2013)
6. W.-J. Beyn, R. Kruse, Two-sided error estimates for the stochastic theta method. Discrete Contin. Dyn. Syst. Ser. B **14**(2), 389–407 (2010)
7. S.C. Brenner, L.R. Scott, *The Mathematical Theory of Finite Element Methods*, 3rd edn. Texts in Applied Mathematics, vol. 15 (Springer, New York, 2008)
8. Z. Brzeźniak, E. Hausenblas, Maximal regularity for stochastic convolutions driven by Lévy processes. Probab. Theory Relat. Fields **145**(3–4), 615–637 (2009)
9. C. Carstensen, Merging the Bramble-Pasciak-Steinbach and the Crouzeix-Thomée criterion for H^1-stability of the L^2-projection onto finite element spaces. Math. Comput. **71**(237), 157–163 (electronic) (2002)
10. C. Carstensen, An adaptive mesh-refining algorithm allowing for an H^1 stable L^2 projection onto Courant finite element spaces. Constr. Approx. **20**(4), 549–564 (2004)
11. K. Chrysafinos, L.S. Hou, Error estimates for semidiscrete finite element approximations of linear and semilinear parabolic equations under minimal regularity assumptions. SIAM J. Numer. Anal. **40**(1), 282–306 (2002)
12. P.G. Ciarlet, *The Finite Element Method for Elliptic Problems*. Classics in Applied Mathematics, vol. 40 (SIAM, Philadelphia, 2002)
13. J.M.C. Clark, R.J. Cameron, The maximum rate of convergence of discrete approximations for stochastic differential equations, in *Stochastic Differential Systems (Proc. IFIP-WG 7/1 Working Conf., Vilnius, 1978)*. Lecture Notes in Control and Information Sciences, vol. 25 (Springer, Berlin, 1980), pp. 162–171
14. D.L. Cohn, *Measure Theory* (Birkhäuser, Boston [u.a.], 1993)
15. S.G. Cox, J. van Neerven, Pathwise Hölder convergence of the implicit Euler scheme for semi-linear SPDEs with multiplicative noise. Numer. Math. **125**(2), 259–345 (2013)
16. M. Crouzeix, V. Thomée, The stability in L_p and W_p^1 of the L_2-projection onto finite element function spaces. Math. Comput. **48**(178), 521–532 (1987)
17. G. Da Prato, A. Lunardi, Maximal regularity for stochastic convolutions in L^p spaces. Atti Accad. Naz. Lincei Cl. Sci. Fis. Mat. Natur. Rend. Lincei (9) Mat. Appl. **9**(1), 25–29 (1998)

R. Kruse, *Strong and Weak Approximation of Semilinear Stochastic Evolution Equations*,   171
Lecture Notes in Mathematics 2093, DOI 10.1007/978-3-319-02231-4,
© Springer International Publishing Switzerland 2014

18. G. Da Prato, J. Zabczyk, *Stochastic Equations in Infinite Dimensions*. Encyclopedia of Mathematics and Its Applications, vol. 44 (Cambridge University Press, Cambridge, 1992)
19. G. Da Prato, S. Kwapien, J. Zabczyk, Regularity of solutions of linear stochastic equations in Hilbert spaces. Stochastics **23**(3), 1–23 (1988)
20. G. Da Prato, A. Jentzen, M. Röckner, A mild Itô formula for SPDEs (2011). Preprint [arXiv:1009.3526v3]
21. A. Debussche, Weak approximation of stochastic partial differential equations: The nonlinear case. Math. Comput. **80**(273), 89–117 (2011)
22. A. Debussche, J. Printems, Weak order for the discretization of the stochastic heat equation. Math. Comput. **78**(266), 845–863 (2009)
23. C.M. Elliott, S. Larsson, Error estimates with smooth and nonsmooth data for a finite element method for the Cahn-Hilliard equation. Math. Comput. **58**(198), 603–630, S33–S36 (1992)
24. K.-J. Engel, R. Nagel, *One-Parameter Semigroups for Linear Evolution Equations*. Graduate Texts in Mathematics, vol. 194 (Springer, Berlin, 2000)
25. L.C. Evans, in *Partial Differential Equations*. Graduate Studies in Mathematics, vol. 19 (American Mathematical Society, Providence, 1998)
26. M. Geissert, M. Kovács, S. Larsson, Rate of weak convergence of the finite element method for the stochastic heat equation with additive noise. BIT Numer. Math. **49**, 343–356 (2009)
27. M.B. Giles, Improved multilevel Monte Carlo convergence using the Milstein scheme, in Monte Carlo and Quasi-Monte Carlo Methods 2006 (Springer, Berlin, 2008), pp. 343–358
28. M.B. Giles, Multilevel Monte Carlo path simulation. Oper. Res. **56**(3), 607–617 (2008)
29. P. Glasserman, *Monte Carlo Methods in Financial Engineering*. Applications of Mathematics (New York), vol. 53. Stochastic Modelling and Applied Probability (Springer, New York, 2004)
30. R.D. Grigorieff, Diskrete Approximation von Eigenwertproblemen. I. Qualitative Konvergenz. Numer. Math. **24**(4), 355–374 (1975)
31. T.H. Gronwall, Note on the derivatives with respect to a parameter of the solutions of a system of differential equations. Ann. Math. (2) **20**(4), 292–296 (1919)
32. A. Grorud, É. Pardoux, Intégrales Hilbertiennes anticipantes par rapport à un processus de Wiener cylindrique et calcul stochastique associé. Appl. Math. Optim. **25**(1), 31–49 (1992)
33. M. Hanke-Bourgeois, *Grundlagen der numerischen Mathematik und des wissenschaftlichen Rechnens*, 2nd edn. Mathematische Leitfäden [Mathematical Textbooks] (B.G. Teubner, Wiesbaden, 2006)
34. E. Hausenblas, Approximation for semilinear stochastic evolution equations. Potential Anal. **18**(2), 141–186 (2003)
35. E. Hausenblas, Weak approximation of the stochastic wave equation. J. Comput. Appl. Math. **235**(1), 33–58 (2010)
36. D. Henry, *Geometric Theory of Semilinear Parabolic Equations*. Lecture Notes in Mathematics, vol. 840 (Springer, Berlin, 1981)
37. J.S. Hesthaven, S. Gottlieb, D. Gottlieb, in *Spectral Methods for Time-Dependent Problems*. Cambridge Monographs on Applied and Computational Mathematics, vol. 21 (Cambridge University Press, Cambridge, 2007)
38. A. Jentzen, P.E. Kloeden, The numerical approximation of stochastic partial differential equations. Milan J. Math. **77**, 205–244 (2009)
39. A. Jentzen, P.E. Kloeden, *Taylor Approximations for Stochastic Partial Differential Equations*. CBMS-NSF Regional Conference Series in Applied Mathematics, vol. 83 (SIAM, Philadelphia, 2011)
40. A. Jentzen, M. Röckner, A Milstein scheme for SPDEs (2012). Preprint [arXiv:1001.2751v4]
41. A. Jentzen, M. Röckner, Regularity analysis for stochastic partial differential equations with nonlinear multiplicative trace class noise. J. Differ. Equ. **252**(1), 114–136 (2012)
42. A. Jentzen, P.E. Kloeden, A. Neuenkirch, Pathwise convergence of numerical schemes for random and stochastic differential equations, in *Foundation of Computational Mathematics*, ed. by F. Cucker, A. Pinkus, M. Todd (Cambridge University Press, Cambridge, 2009), pp. 140–161 (Hong Kong, 2008)

43. A. Jentzen, P.E. Kloeden, A. Neuenkirch, Pathwise approximation of stochastic differential equations on domains: Higher order convergence rates without global Lipschitz coefficients. Numer. Math. **112**(1), 41–64 (2009)
44. M. Kovács, S. Larsson, F. Lindgren, Strong convergence of the finite element method with truncated noise for semilinear parabolic stochastic equations with additive noise. Numer. Algorithms **53**(2–3), 309–320 (2010)
45. M. Kovács, S. Larsson, F. Lindgren, Weak convergence of finite element approximations of linear stochastic evolution equations with additive noise. BIT Numer. Math. **52**(1), 85–108 (2012)
46. M. Kovács, S. Larsson, F. Lindgren, Weak convergence of finite element approximations of linear stochastic evolution equations with additive noise II. Fully discrete schemes. BIT Numer. Math. **53**(2), 497–525 (2012). Preprint [arXiv:1203.2029v1]
47. R. Kruse, Characterization of bistability for stochastic multistep methods. BIT Numer. Math. **52**(1), 109–140 (2012)
48. R. Kruse, Consistency and stability of a Milstein-Galerkin finite element scheme for semilinear SPDE (2013). ArXiv Preprint [arXiv:1307.4120v1] (submitted)
49. R. Kruse, Optimal error estimates of Galerkin finite element methods for stochastic partial differential equations with multiplicative noise. IMA J. Numer. Anal. (2013) (Online First)
50. R. Kruse, S. Larsson, Optimal regularity for semilinear stochastic partial differential equations with multiplicative noise. Electron. J. Probab. **17**(65), 1–19 (2012)
51. A. Lang, P.-L. Chow, J. Potthoff, Almost sure convergence of a semidiscrete Milstein scheme for SPDEs of Zakai type. Stochastics **82**(3), 315–326 (2010)
52. S. Larsson, Semilinear parabolic partial differential equations: Theory, approximation, and application, in *New Trends in the Mathematical and Computer Sciences*. Publ. ICMCS, vol. 3 (Int. Cent. Math. Comp. Sci. (ICMCS), Lagos, 2006), pp. 153–194
53. S. Larsson, V. Thomée, *Partial Differential Equations with Numerical Methods*. Texts in Applied Mathematics, vol. 45 (Springer, Berlin, 2003)
54. J.A. León, D. Nualart, Stochastic evolution equations with random generators. Ann. Probab. **26**(1), 149–186 (1998)
55. P. Malliavin, Stochastic calculus of variation and hypoelliptic operators, in *Proceedings of the International Symposium on Stochastic Differential Equations (Res. Inst. Math. Sci., Kyoto Univ., Kyoto, 1976)* (Wiley, New York, 1978), pp. 195–263
56. X. Mao, *Stochastic Differential Equations and Their Applications*. Horwood Publishing Series in Mathematics & Applications (Horwood Publishing Limited, Chichester, 1997)
57. M. Miklavčič, *Applied Functional Analysis and Partial Differential Equations* (World Scientific, River Edge, 1998)
58. D. Nualart, *The Malliavin Calculus and Related Topics*, 2nd edn. Probability and Its Applications (New York) (Springer, Berlin, 2006)
59. D. Nualart, Application of Malliavin calculus to stochastic partial differential equations, in *A Minicourse on Stochastic Partial Differential Equations*. Lecture Notes in Mathematics, vol. 1962 (Springer, Berlin, 2009), pp. 73–109
60. A. Pazy, *Semigroups of Linear Operators and Applications to Partial Differential Equations*. Applied Mathematical Sciences, vol. 44 (Springer, New York, 1983)
61. C. Prévôt, M. Röckner, *A Concise Course on Stochastic Partial Differential Equations*. Lecture Notes in Mathematics, vol. 1905 (Springer, Berlin, 2007)
62. M. Reed, B. Simon, *Methods of Modern Mathematical Physics*, 2nd edn. Functional Analysis, vol. 1 (Academic [Harcourt Brace Jovanovich, Publishers], New York, 1980), xv+400 pp. ISBN:0-12-585050-6
63. M. Röckner, Z. Sobol, Kolmogorov equations in infinite dimensions: Well-posedness and regularity of solutions, with applications to stochastic generalized Burgers equations. Ann. Probab. **34**(2), 663–727 (2006)
64. M. Röckner, Z. Sobol, A new approach to Kolmogorov equations in infinite dimensions and applications to the stochastic 2D Navier-Stokes equation. C. R. Math. Acad. Sci. Paris **345**(5), 289–292 (2007)

65. G.R. Sell, Y. You, *Dynamics of Evolutionary Equations*, vol. 143 (Springer, New York, 2002)
66. V. Thomée, *Galerkin Finite Element Methods for Parabolic Problems*, 2nd edn. Springer Series in Computational Mathematics, vol. 25 (Springer, Berlin, 2006)
67. J. van Neerven, M. Veraar, L. Weis, Stochastic maximal L^p-regularity. SIAM J. Math. Anal. **44**(3), 1372–1414 (2012)
68. M. Veraar, The stochastic Fubini theorem revisited. Stochastics **84**(4), 543–551 (2012)
69. D. Werner, *Funktionalanalysis*, 5th extended edn. (Springer, Berlin, 2004)
70. Y. Yan, Galerkin finite element methods for stochastic parabolic partial differential equations. SIAM J. Numer. Anal. **43**(4), 1363–1384 (2005)

Index

R. Kruse, *Strong and Weak Approximation of Semilinear Stochastic Evolution Equations*,
Lecture Notes in Mathematics 2093, DOI 10.1007/978-3-319-02231-4,
© Springer International Publishing Switzerland 2014

LECTURE NOTES IN MATHEMATICS   🐎 Springer

Edited by J.-M. Morel, B. Teissier; P.K. Maini

Editorial Policy (for the publication of monographs)

1. Lecture Notes aim to report new developments in all areas of mathematics and their applications - quickly, informally and at a high level. Mathematical texts analysing new developments in modelling and numerical simulation are welcome.

   Monograph manuscripts should be reasonably self-contained and rounded off. Thus they may, and often will, present not only results of the author but also related work by other people. They may be based on specialised lecture courses. Furthermore, the manuscripts should provide sufficient motivation, examples and applications. This clearly distinguishes Lecture Notes from journal articles or technical reports which normally are very concise. Articles intended for a journal but too long to be accepted by most journals, usually do not have this "lecture notes" character. For similar reasons it is unusual for doctoral theses to be accepted for the Lecture Notes series, though habilitation theses may be appropriate.

2. Manuscripts should be submitted either online at www.editorialmanager.com/lnm to Springer's mathematics editorial in Heidelberg, or to one of the series editors. In general, manuscripts will be sent out to 2 external referees for evaluation. If a decision cannot yet be reached on the basis of the first 2 reports, further referees may be contacted: The author will be informed of this. A final decision to publish can be made only on the basis of the complete manuscript, however a refereeing process leading to a preliminary decision can be based on a pre-final or incomplete manuscript. The strict minimum amount of material that will be considered should include a detailed outline describing the planned contents of each chapter, a bibliography and several sample chapters.

   Authors should be aware that incomplete or insufficiently close to final manuscripts almost always result in longer refereeing times and nevertheless unclear referees' recommendations, making further refereeing of a final draft necessary.

   Authors should also be aware that parallel submission of their manuscript to another publisher while under consideration for LNM will in general lead to immediate rejection.

3. Manuscripts should in general be submitted in English. Final manuscripts should contain at least 100 pages of mathematical text and should always include

   – a table of contents;
   – an informative introduction, with adequate motivation and perhaps some historical remarks: it should be accessible to a reader not intimately familiar with the topic treated;
   – a subject index: as a rule this is genuinely helpful for the reader.

   For evaluation purposes, manuscripts may be submitted in print or electronic form (print form is still preferred by most referees), in the latter case preferably as pdf- or zipped ps-files. Lecture Notes volumes are, as a rule, printed digitally from the authors' files. To ensure best results, authors are asked to use the LaTeX2e style files available from Springer's web-server at:

   ftp://ftp.springer.de/pub/tex/latex/svmonot1/ (for monographs) and
   ftp://ftp.springer.de/pub/tex/latex/svmultt1/ (for summer schools/tutorials).

Additional technical instructions, if necessary, are available on request from lnm@springer.com.

4. Careful preparation of the manuscripts will help keep production time short besides ensuring satisfactory appearance of the finished book in print and online. After acceptance of the manuscript authors will be asked to prepare the final LaTeX source files and also the corresponding dvi-, pdf- or zipped ps-file. The LaTeX source files are essential for producing the full-text online version of the book (see http://www.springerlink.com/openurl.asp?genre=journal&issn=0075-8434 for the existing online volumes of LNM). The actual production of a Lecture Notes volume takes approximately 12 weeks.

5. Authors receive a total of 50 free copies of their volume, but no royalties. They are entitled to a discount of 33.3 % on the price of Springer books purchased for their personal use, if ordering directly from Springer.

6. Commitment to publish is made by letter of intent rather than by signing a formal contract. Springer-Verlag secures the copyright for each volume. Authors are free to reuse material contained in their LNM volumes in later publications: a brief written (or e-mail) request for formal permission is sufficient.

Addresses:
Professor J.-M. Morel, CMLA,
École Normale Supérieure de Cachan,
61 Avenue du Président Wilson, 94235 Cachan Cedex, France
E-mail: morel@cmla.ens-cachan.fr

Professor B. Teissier, Institut Mathématique de Jussieu,
UMR 7586 du CNRS, Équipe "Géométrie et Dynamique",
175 rue du Chevaleret
75013 Paris, France
E-mail: teissier@math.jussieu.fr

For the "Mathematical Biosciences Subseries" of LNM:

Professor P. K. Maini, Center for Mathematical Biology,
Mathematical Institute, 24-29 St Giles,
Oxford OX1 3LP, UK
E-mail: maini@maths.ox.ac.uk

Springer, Mathematics Editorial, Tiergartenstr. 17,
69121 Heidelberg, Germany,
Tel.: +49 (6221) 4876-8259

Fax: +49 (6221) 4876-8259
E-mail: lnm@springer.com